Lecture Notes
in Economics and
Mathematical Systems

Managing Editors: M. Beckmann and H. P. Künzi

Econometrics

100

Benjamin S. Duran
Patrick L. Odell

Cluster Analysis
A Survey

Springer-Verlag
Berlin · Heidelberg · New York 1974

Managing Editors
Prof. Dr. M. Beckmann
Brown University
Providence, RI 02912/USA

Prof. Dr. H. P. Künzi
Universität Zürich
8090 Zürich/Schweiz

Dr. Benjamin S. Duran
Texas Tech University
Dept. of Mathematics
Lubbock, Texas 79409/USA

Dr. Patrick L. Odell
The University of Texas at Dallas
Dallas, Texas 75230/USA

Library of Congress Cataloging in Publication Data

Duran, Benjamin S   1939-
    Cluster analysis.

    (Lecture notes in economics and mathematical
systems, 100. Operations research)
    Bibliography: p.
    1. Cluster analysis. I. Odell, Patrick L.,
1930-   joint author. II. Title. III. Series:
Lecture notes in economics and mathematical systems, 100.
IV. Series: Poerations research.
QA278.D87   519.5'3   74-17099

AMS Subject Classifications (1970): 62 H 99, 62 P 20

ISBN 3-540-06954-2 Springer-Verlag Berlin · Heidelberg · New York
ISBN 0-387-06954-2 Springer-Verlag New York · Heidelberg · Berlin

PREFACE

A tremendous amount of work has been done over the last thirty years in cluster analysis, with a significant amount occurring since 1960. A substantial portion of this work has appeared in many journals, including numerous applied journals, and a unified exposition is lacking.

The purpose of this monograph is to supply such an exposition by presenting a brief survey on cluster analysis. The main intent of the monograph is to give the reader a quick account of the problem of cluster analysis and to expose to him the various aspects thereof. With this intent in mind much detail has been omitted, particularly in so far as detailed examples are considered. Most of the references stated within the text contain examples and the reader can consult them for additional information on specific topics. Efforts were made to include in the reference section all papers that played a role in developing the "theory" of cluster analysis. Any omission of such references was not intentional and we would appreciate knowing about them. Many references to papers in applied journals are also contained, however, the list is far from being complete.

This monograph has been greatly influenced by the work of many people, most notably, J.A. Hartigan, D. Wishart, J.K. Bryan, R.E. Jensen, H.D. Vinod, and M.R. Rao.

Several portions of the monograph were motivated by research performed under the support of NASA Manned Spacecraft Center, Earth Observations Division, under Contract NAS 9-12775.

TABLE OF CONTENTS

Chapter 1

THE CLUSTER PROBLEM AND PRELIMINARY IDEAS

1.1 Basic Notions and Definitions

There exist many situations in scientific and business investigations in which the technique which has come to be called cluster analysis is applicable. In order to avoid specialization to perhaps a few fields of application we will discuss the technique from a general and perhaps somewhat abstract approach in this chapter and give what we consider interesting applications in a later chapter.

Let the set $I = \{I_1, I_2, \ldots, I_n\}$ denote n individuals from a conceptual population π_I. It is tacitly assumed that there exists a set of features or characteristics $C = (C_1, C_2, \ldots, C_p)^T$ which are observable and are possessed by each individual in I. The term observable is used here to denote characteristics that yield both quantitative and qualitative data; although we will base most of our discussion on quantitative data which we will at times call measurements. We denote the value of the measurement on the ith characteristic of the individual I_j by the symbol x_{ij} and let $X_j = [x_{ij}]$ denote the p x 1 vector of such measurements. Hence for a set of individuals I, there is available to the investigator a corresponding set of p x 1 measurement vectors $X = \{X_1, X_2, \ldots, X_n\}$ which describe the set I. It is important to note that the set X can be thought of as n points in p-dimensional Euclidean space, E_p.

1.2. The Cluster Problem

Let m be an integer less than n. Based on the data contained in the set X the cluster problem is to determine m clusters (subsets) of individuals in I, say $\pi_1, \pi_2, \ldots, \pi_m$, such that I_i belongs to one and only one subset and those individuals which are assigned to the same cluster are similar yet individuals from different clusters are different (not similar).

A solution to the cluster problem is usually to determine a partitioning that satisfies some optimality criterion. This optimality criterion may be given in terms of a functional relation that reflects the levels of desirability of the various partitions or groupings. This functional relation is often called an <u>objective function</u>. For example, the within group sum of squares [see section 1.5] may be used as an objective function. As an example suppose one characteristic (i.e., p = 1) is measured on each of n = 8 individuals resulting in the set X = {3,4,7,4,3,3,4,4 }. The within group sum of squares is given by the functional relation,

$$W = \sum_{i=1}^{n} (x_i - \bar{x})^2 = \sum_{i=1}^{n} x_i^2 - \frac{1}{n} (\sum_{i=1}^{n} x_i)^2$$

where x_i represents the measurement on the ith individual. For the group containing all 8 individuals the functional relation yields

$$\sum_{i=1}^{8} x_i^2 - \frac{1}{8} (\sum_{i=1}^{8} x_i)^2 = 140 - 128 = 12.$$

If the set X is partitioned into the three groups, $G_1 = \{3,3,3\}$, $G_2 = \{4,4,4,4\}$, and $G_3 = \{7\}$, then the within groups sum of squares becomes

$$W_1 + W_2 + W_3 = 0 + 0 + 0 = 0$$

where W_i denotes the sum of squares corresponding to G_i. The optimal value in this example is 0, if one desires three groups. In general one must consider both, the values of the objective function and the number of groups desired. Various types of objective functions will be defined, many of which can be formulated in a unified and general manner.

In order to "solve" the cluster problem it is clearly desirable to define the terms similarity and difference in a quantitative fashion. What does it mean to say two individuals I_j and I_k are different? A solution to the problem could follow perhaps if one would assign the ith and j^{th} individuals to the same cluster if the

distance between the points X_i and X_j is "sufficiently small" and
to different clusters if the distance between X_i and X_j is "sufficiently
large". Hence it is germane to our purpose to review what is meant
abstractly by a distance between the points X_i and X_j in E_p.

1.3. Distance Functions

Definition 1.1. A non-negative real valued function
$d(X_i, X_j)$ is said to be a distance function (metric) if

(a) $d(X_i, X_j) \geq 0$ for all X_i and X_j in E_p,

(b) $d(X_i, X_j) = 0$ if and only if $X_i = X_j$,

(c) $d(X_i, X_j) = d(X_j, X_i)$,

(d) $d(X_i, X_j) \leq d(X_i, X_k) + d(X_k, X_j)$,

where X_i, X_j and X_k are any three vectors in E_p.

The value of $d(X_i, X_j)$ for specified X_i and X_j is said to be
the distance between X_i and X_j and equivalently the distance between
I_i and I_j with respect to the selected characteristics $C = (C_1, C_2, \ldots, C_p)^T$.

Examples of some popular and useful distance functions are given
in Table 1.1.

TABLE 1.1

Some Distance Functions

NAME	FORM
1. Euclidean	$d_2(X_i, X_j) = [\sum_{k=1}^{p} (x_{ki} - x_{kj})^2]^{1/2}$
2. ℓ_1 norm	$d_1(X_i, X_j) = [\sum_{k=1}^{p} |x_{ki} - x_{kj}|]$
3. Sup-norm	$d_\infty(X_i, X_j) = \sup_{k=1,2,\ldots,p} \{|x_{ki} - x_{kj}|\}$
4. ℓ_p norm	$d_p(X_i, X_j) = [\sum_{k=1}^{p} |x_{ki} - x_{kj}|^p]^{1/p}$
5. Mahalanobis [252]	$D^2(X_i, X_j) = (X_i - X_j)^T W^{-1} (X_i - X_j)$

The Euclidean metric is a very popular and commonly used metric. The absolute value metric is easy to evaluate computationally. The sup-norm is also simple computationally; however does involve a ranking procedure. The ℓ_p norm includes distance functions 1, 2, and 3 as special cases for $p = 2$, 1, and ∞, respectively.

The Mahalanobis metric is often referred to as generalized Euclidean distance. The matrix W^{-1} usually denotes the inverse of the within scatter matrix (see section 1.5). The Mahalanobis distance is invariant under any nonsingular linear transformation. Consider the transformation $Y = BX$. Then

$$
\begin{aligned}
D^2(Y_i, Y_j) &= (Y_i - Y_j)^T W_y^{-1} (Y_i - Y_j) \\
&= (BX_i - BX_j)^T W_y^{-1} (BX_i - BX_j) \\
&= (X_i - X_j)^T B^T W_y^{-1} B (X_i - X_j) \\
&= (X_i - X_j)^T B^T (BW_x B^T)^{-1} B (X_i - X_j) \\
&= (X_i - X_j)^T W_x^{-1} (X_i - X_j) \\
&= D^2(X_i, X_j)
\end{aligned}
$$

There are other heuristic measures of distance which are not true metrics yet have been used. For example, the Jeffreys-Matusita [81], [82], [259] measure of distance which is given by

$$
M = [\sum_{k=1}^{p} (\sqrt{x_{ki}} - \sqrt{x_{kj}})^2]^{1/2} \tag{1.1}
$$

and a measure of distance called the coefficient of divergence [55] given by

$$
CD = \{\frac{1}{p} \sum_{k=1}^{p} (\frac{x_{ki} - x_{kj}}{x_{ki} + x_{kj}})^2\}^{1/2} .
$$

The Jeffreys-Matusita measure was originally defined as a distance between two probability density functions, however, in the form (1.1) it can be used as a measure of distance between a pair of vectors.

In the original use of the coefficient of divergence the x's were
actually means $\bar{x}$'s, and one was considering the distance between the
sample means of two samples.

The following theorem gives an ordering of the distance
functions defined by the ℓ_p norm.

Theorem 1.1.   The inequality

$$d_h(X_i,X_j) \leq d_m(X_i,X_j)$$

holds for all X_i and X_j in E_p if and only if $h \geq m$.

However, one recalls that the definition of distance requires
$X_i = X_j$ for $d_p(X_i,X_j) = 0$.

1.4.   Measures of Similarity

The n measurements $X_1, X_2, \ldots, X_n$ may be represented in terms of
the p x n data matrix

$$X = \begin{pmatrix} x_{11} & x_{12} & \cdots & x_{1n} \\ x_{21} & x_{22} & \cdots & x_{2n} \\ \vdots & \vdots & & \\ x_{p1} & x_{p2} & \cdots & x_{pn} \end{pmatrix} = (X_1, X_2, \ldots, X_n).$$

Likewise the pairwise distances $d(X_i,X_j)$ may be represented in terms
of the symmetric n x n distance matrix

$$D = \begin{pmatrix} 0 & d_{12} & \cdots & d_{1n} \\ d_{21} & 0 & \cdots & d_{2n} \\ \vdots & \vdots & & \\ d_{n1} & d_{n2} & \cdots & 0 \end{pmatrix}$$

Note that the diagonal elements of D are $d_{ii} = 0$ for $i = 1,2,\ldots,n$.

As a complement to the notion of distance between X_i and X_j is
the idea of similarity between two individuals I_i and I_j.

<u>Definition 1.2</u>. A non-negative real valued function $s(X_i, X_j) = s_{ij}$ is said to be a similarity measure if

$\quad$ (i) $\quad 0 \leq s(X_i, X_j) < 1$ for $X_i \neq X_j$.

$\quad$ (ii) $\quad s(X_i, X_i) = 1$,

$\quad$ (iii) $\quad s(X_i, X_j) = s(X_j, X_i)$.

$\quad$ The pairwise similarities may be arranged in the similarity matrix

$$S = \begin{pmatrix} 1 & s_{12} & \cdots & s_{1n} \\ s_{21} & 1 & \cdots & s_{2n} \\ s_{n1} & s_{n2} & \cdots & 1 \end{pmatrix}.$$

The quantity s_{ij} will be called simply a similarity coefficient. It is also known as a coefficient of association or a matching coefficient when each measurement vector X_i contains only 0's and 1's.

$\quad$ There are some coefficients of association which range in value between -1 and 1, such as the phi coefficient, known also as the fourfold point correlation coefficient. However, we will confine our study to the coefficient of definition 1.2.

$\quad$ Suppose that each observation vector contains only 0's and 1's, that is, we have binary data. Given two measurement vectors X_i and X_j define n_{IJ} to be the number of characteristics which yield 1 in both X_i and X_j; n_{ij} the number that yield 0 in both; n_{iJ} the number that yield 0 in X_i and 1 in X_j; and similarly for n_{Ij}. Thus $n_J = n_{IJ} + n_{iJ}$ is the number of 1's in X_j and $n_j = n_{Ij} + n_{ij}$ is the number of 0's in X_j. Table 1.2 contains a list of some similarity coefficients in terms of the quantities defined here. A discussion of the similarity coefficients in Table 1.2 and various other coefficients is contained in [336].

TABLE 1.2

Some Similarity Coefficients

For Binary Data

Coefficient	References
$\dfrac{n_{IJ}}{n_{IJ} + n_{Ij} + n_{iJ}}$	[173], [328]
$\dfrac{n_{IJ} + n_{ij}}{p}$	[334]
$\dfrac{n_{IJ}}{p}$	[303]
$\dfrac{2n_{IJ}}{2n_{IJ} + n_{Ij} + n_{iJ}}$	[82], [338]
$\dfrac{2(n_{IJ} + n_{ij})}{p + n_{IJ} + n_{ij}}$	
$\dfrac{n_{IJ}}{n_{IJ} + 2(n_{Ij} + n_{iJ})}$	
$\dfrac{n_{IJ} + n_{ij}}{p + n_{Ij} + n_{iJ}}$	[294]

Statisticians have historically used a measure of linear similarity called the correlation coefficient, usually denoted by r_{ij}, where

$$r_{ij} = [\sum_{k=1}^{p} X_{ki}X_{kj}] / [\sum_{k=1}^{p} X_{ki}^2 \sum_{k=1}^{p} X_{kj}^2]^{1/2}. \qquad (1.2)$$

Written in the form (1.2) it is assumed that $\sum_{k=1}^{p} X_{ki} = \sum_{k=1}^{p} X_{kj} = 0$. This quantity r_{ij} holds an esteem place in statistics and is used and misused frequently by almost everyone. It is important to note that if one considers X_i and X_j the coordinates of two points in E_p and end points of two vectors both originating at the origin then [7]

$$r_{ij} = \cos \theta \qquad (1.3)$$

where θ is the angle between the two vectors. Hence, r_{ij} is such that $-1 < r_{ij} < 1$, and individuals I_i and I_j are said to be similar in a positive way if r_{ij} is "close to" unity; similar in a negative way if r_{ij} is "close to" negative unity; and not similar if r_{ij} is "close to" zero. Note that r_{ij} is not a true similarity function as we have defined that term.

Lemma 1.1. The value of $r_{ij} = 1$ if and only if $X_i = kX_j$ where k is any non-negative scalar.

The proof of this lemma follows directly from the definition in (1.2). Note that two points X_1 and X_2 may be relatively distant from each other yet have similarity 1 with respect to the measure r_{ij}. Consider the numerical example illustrated by Fig. 1.

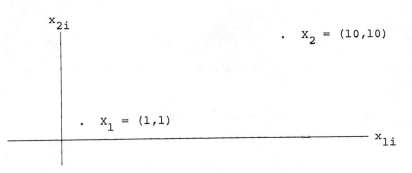

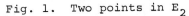

Fig. 1. Two points in E_2

Using metrics (1), (2), (3) in Table 1.1 and r_{ij} in equation (1.2) one computes

$$d_2(X_1,X_2) = [(10 - 1)^2 + (10 - 1)^2]^{1/2} = 9\sqrt{2}$$

$$d_1(X_1,X_2) = [|10 - 1| + |10 - 1|] = 18$$

$$d_\infty(X_1,X_2) = \sup [(10 - 1), (10 - 1)] = 9$$

and

$$r_{12} = 1$$

Note that even though $X_1 \neq X_2$, that $r_{ij} = 1$ implies that the individuals I_1 and I_2 will be judged similar. Note also

$$d_\infty(X_1,X_2) < d_2(X_1,X_2) < d_1(X_1,X_2)$$

which illustrates Theorem 1.1 of section 1.3.

It is important to note that one can use the various distance measures in section 1.3 to construct corresponding similarity measures by judiciously selecting an appropriate transformation. Thus, if one prefers to work with similarities rather than distances he may by performing the necessary modifications.

Let us now use our concept of distance for computing a measure of scatter or inhomogeneity of a set of individuals $I = \{I_1,...,I_n\}$.
Definition 1.3. Let $X = \{X_1,X_2,...,X_n\}$ denote the set of observations made on the set of individuals $I = \{I_1,I_2,...,I_n\}$. Then the scalar s_d where

$$s_d = \frac{1}{2} \sum_{i=1}^{n} \sum_{j=1}^{n} d(X_i, X_j) \tag{1.4}$$

is said to be the <u>total</u> scatter with respect to a distance function $d(X_i, X_j)$.

<u>Definition 1.4.</u> The value $\overline{s_d} = s_d/N_d$ where $N_d = (n^2 - n)/2$ is said to be the mean scatter of the set I.

The rationale for the Definitions 1.3 and 1.4 follow easily by considering the distance matrix $D = \{d_{ij} = d(X_i, X_j)\}$ and noting that

$$d_{ii} = d(X_i, X_i) = 0 \text{ for all } i$$

$$d(X_i, X_j) = d(X_j, X_i) \text{ implies } d_{ij} = d_{ji} \text{ for all } i \neq j = 1, 2, \ldots, n.$$

This implies that there are n^2 distance involved in computing s_d of which n are always zero and $(n^2 - n)/2$ are in general distinct and non-negative. Hence $\overline{s_d}$ is the arithmetic average of the non-zero distinct distances associated with the set X or equivalently with the set I. The matrix D gives a compact way of displaying the distances among the elements of the set I.

The statisticians have used a similar measure for scatter and we introduce that here [Wilks, 591-614].

<u>Definition 1.5.</u> The p x p matrix

$$S_x = \sum_{i=1}^{n} (X_i - \overline{X})(X_i - \overline{X})^T \tag{1.5}$$

is called the <u>scatter</u> <u>matrix</u> for the set X, where

$$\overline{X} = \sum_{i=1}^{n} X_i/n \tag{1.6}$$

is a p x 1 vector of arithmetic averages.

The matrix S_x is sometimes called the <u>matrix</u> <u>sum</u> <u>of</u> <u>squares</u>.

Definition 1.6. The trace of S_x is said to be the __statistical scatter__ of the set X and is denoted by

$$s_t = \text{tr } S_x = \sum_{i=1}^{n} \sum_{k=1}^{p} (X_{ki} - \bar{X}_k)^2 = \sum_{i=1}^{n} (X_i - \bar{X})^T (X_i - \bar{X}).$$

The measure s_t is the sum of the distances of the n points from the group mean $\bar{X}$ and is termed the error (within) sum of squares. It can be shown that

$$s_t = \frac{1}{n} \sum_{\substack{i=1 \\ i<j}}^{n} \sum_{j=1}^{n} (X_i - X_j)^T (X_i - X_j)$$

$$= \frac{1}{n} \sum_{\substack{i=1 \\ i<j}}^{n} \sum_{j=1}^{n} d^2(X_i, X_j) . \tag{1.7}$$

Thus when dealing with tr S_x one is implicitly using Euclidean distance.

Definition 1.7. The determinant $|S_x|$, of the matrix S_x is said to be the statistical scatter with respect to the determinant and is denoted by $s_D = |S_x|$.

The matrix of correlation coefficients $R = \{r_{ij}\}$ can be computed from the matrix $S_x = \{s_{ij}\}$ defined by (1.5). Define the diagonal matrices, $[\text{Dia } S_x] = \{s_{11}, s_{22}, \dots, s_{pp}\}$ and $[\text{Dia } S_x]^{1/2} = \{s_{11}^{1/2}, s_{22}^{1/2}, \dots, s_{pp}^{1/2}\}$. Then

$$R = [\text{Dia } S_x]^{1/2} S_x [\text{Dia } S_x]^{1/2}. \tag{1.8}$$

Lemma 1.2. $S_x = \phi$, the null matrix, if and only if X_1 and $X_2 = \dots = X_k$ and $X_{k+1} = \dots = X_n = \phi$ for some $k \le n$.

1.5. Distance and Similarity Between Clusters

Many clustering procedures, as will be seen later, are

hierarchichal. That is, the two closest objects I_1 and I_2 are grouped and treated as a single cluster. Thus, the number of objects is reduced to $n - 1$, a single cluster of 2 objects and $n - 2$ clusters of a single object in each, and the process is repeated until the objects have been grouped into the cluster containing all n objects. Implicit in this hierarchical process is the concept of distance between an object and a cluster and distance between two clusters.

Inherent in the statement of the cluster problem is the concept of optimality criterion (objective function) which dictates when a desirable partitioning has been obtained. Thus we need a measure of homogeneity within a cluster and disparity between two clusters. These two measures can very well depend on the distance between two clusters.

Let $I = \{I_1, I_2, \ldots, I_{n_1}\}$ and $J = \{J_1, J_2, \ldots, J_{n_2}\}$ denote two clusters of individuals from a population Π. Let $C = (C_1, C_2, \ldots, C_p)^T$ be a set of characteristics which generate the two measurement sets $X = \{X_1, X_2, \ldots, X_{n_1}\}$ and $Y = \{Y_1, Y_2, \ldots, Y_{n_2}\}$, associated with I and J, respectively.

Definition 1.8. Let $\mathcal{D} = \{d(X_i, Y_j)\ i = 2, \ldots, n_1;\ j = 1, 2, \ldots, n_2\}$ denote the set of distances. Then the value

$$D_1(I,J) = \min_{\substack{i=1,\ldots,n_1 \\ j=1,\ldots,n_2}} d(X_i, X_j)$$

is the nearest neighbor distance [395] of I and J with respect to the distance function d.

Definition 1.9. Let $\mathcal{D} = \{d(X_i, Y_j)\}$ be defined as in Definition 1.8, then

$$D_2 = \max_{\substack{i=1,\ldots,n_1 \\ j=1,\ldots,n_2}} d(X_i, Y_j)$$

is the furthest neighbor distance [234] of I and J.

Definition 1.10. The value

$$D_3 = \sum_{j=1}^{n_2} \sum_{i=1}^{n_1} d(X_i, Y_j)/n_1 n_2$$

is the average distance [225] between I and J with respect to the distance function d.

Those who use the concept of statistical scatter sometimes use the following measure of distance between the clusters I and J.

Definition 1.11. The value

$$D_4 = \frac{n_1 n_2}{n_1 + n_2} (\bar{X} - \bar{Y})^T (\bar{X} - \bar{Y})$$

where

$$\bar{X} = \sum_{i=1}^{n_1} X_i/n_1$$

$$\bar{Y} = \sum_{i=1}^{n_2} Y_i/n_2$$

is said to be the <u>statistical</u> <u>distance</u> <u>between</u> <u>the</u> <u>clusters</u> I and J.

The rationale for the measure D_4 arises naturally from the fact that if one considers the clusters I and J as composing a single cluster K where K = I $\cup$ J ($\cup$ denotes "union") then by (1.5)

$$S_K = \sum_{i=1}^{n_1} (X_i - M)(X_i - M)^T + \sum_{i=1}^{n_2} (Y_i - M)(Y_i - M)^T$$

where $M = (\sum_{i=1}^{n_1} X_i + \sum_{i=1}^{n_2} Y_i)/(n_1 + n_2).$

Hence

$$S_K = \sum_{i=1}^{n_1} (X_i - \bar{X} + \bar{X} - M)(X_i - \bar{X} + \bar{X} - M)^T$$

$$+ \sum_{i=1}^{n_2} (Y_i - \bar{Y} + \bar{Y} - M)(Y_i - \bar{Y} + \bar{Y} - M)^T$$

$$= \sum_{i=1}^{n_1} (X_i - \bar{X})(X_i - \bar{X})^T + \sum_{i=1}^{n_1} (\bar{X} - M)(\bar{X} - M)^T$$

$$+ \sum_{i=1}^{n_2} (Y_i - \bar{Y})(Y_i - \bar{Y})^T + \sum_{i=1}^{n_2} (\bar{Y} - M)(\bar{Y} - M)^T$$

since $\sum_{i=1}^{n_1} (X_i - \bar{X})(\bar{X} - M)^T = \sum_{i=1}^{n_2} (Y_i - \bar{Y})(\bar{Y} - M)^T = \phi.$

Note that since

$$M = (n_1\bar{X} + n_2\bar{Y})/(n_1 + n_2)$$

that

$$\sum_{i=1}^{n_1} (\bar{X} - M)(\bar{X} - M)^T = \frac{n_1 n_2^2}{(n_1+n_2)^2} (\bar{X} - \bar{Y})(\bar{X} - \bar{Y})^T$$

and

$$\sum_{i=1}^{n_2} (\bar{Y} - M)(\bar{Y} - M)^T = \frac{n_1^2 n_2}{(n_1+n_2)^2} (\bar{X} - \bar{Y})(\bar{X} - \bar{Y})^T.$$

Hence

$$\sum_{i=1}^{n_1} (\bar{X} - M)(\bar{X} - M)^T + \sum_{i=1}^{n_2} (\bar{Y} - M)(\bar{Y} - M)^T$$

$$= \left[\frac{n_1 n_2^2}{(n_1+n_2)^2} + \frac{n_1^2 n_2}{(n_1+n_2)^2} \right] (\bar{X} - \bar{Y})(\bar{X} - \bar{Y})^T$$

$$= \frac{n_1 n_2}{n_1 + n_2} (\bar{X} - \bar{Y})(\bar{X} - \bar{Y})^T$$

which we call the between scatter matrix.
Finally one can write

$$S_K = S_I + S_J + \frac{n_1 n_2}{n_1 + n_2} (\bar{X} - \bar{Y})(\bar{X} - \bar{Y})^T \qquad (1.9)$$

where S_I and S_J denote the within scatter of I and J, respectively.
The term

$$\frac{n_1 n_2}{n_1 + n_2} (\bar{X} - \bar{Y})(\bar{X} - \bar{Y})^T \qquad (1.10)$$

has been called the between scatter matrix and the trace of this
matrix,

$$\text{tr} \frac{n_1 n_2}{n_1 + n_2} (\bar{X} - \bar{Y})(\bar{X} - \bar{Y})^T = \frac{n_1 n_2}{n_1 + n_2} (\bar{X} - \bar{Y})^T (\bar{X} - \bar{Y})$$

is the statistical distance between the clusters I and J. The
trace of matrix (1.10) is called the within group sum of squares
(WGSS) and is actually the increase in WGSS when the clusters I and
J are combined to form cluster K.

The statistician interprets (1.9) as "the total sum of squares
is equal to the within sum of squares plus the between sum of
squares". The sum $S_I + S_J$ is the "within sum of squares", and (1.10)
gives the "between sum of squares" in matrix form. Clearly all of
the distance measures defined in section 1.3 may be incorporated
into the distance measures between two clusters stated above.

Now, consider each cluster of points as a sample from some
population. Let f and g denote the probability density functions

corresponding to the clusters I and J. Wacker and Langrebe [353] discuss several multivariate forms of distance measures and their metric properties. Their results are summarized in Table 1.3 where C denotes the class of all p-variate absolutely continuous distribution functions, MVN the class of multivariate normal distribution functions, and MVN_Σ the class of multivariate normal distribution functions with equal covariance matrices. The table also gives the metric properties of the distance measures relative to these three classes of distribution functions.

These inter-cluster distance measures might prove useful in those cases where normality can be assumed. In those cases one could conceivably estimate μ and Σ with $\bar{X}$ and s^2 and thereby be able to compute the measures in Table 1.3. The Divergence measure has been used in the application of discriminate analysis to remote sensing data [226]. The Jeffreys-Matusita measure has been used in the application of discriminate analysis to agricultural data in "per field" classification [169].

Most of the reference cited in Table 1.3 deal with univariate forms of the distance measures. Wacker and Langrebe [383] discuss the distance measures in Table 1.3 and their extension to the multivariate case. The reader is referred to [383] for a more extensive discussion of the measures in Table 1.3.

TABLE 1.3

Multivariate Forms of Distance Measures and Their Metric Properties

Name	Form	Metric in C	Metric in MVN	Metric in MVN$_\Sigma$	References
1. Cramer-Von Mises	$W = \{\int_{-\infty}^{\infty} (G(x) - F(x))^2 dx\}^{1/2}$	Yes	Yes	Yes	[67] [381] [76] [304]
2. Kolmogorov-Smirnov	$K = \sup_x \|G(x) - F(x)\|$	Yes	Yes	Yes	[207] [326] [76] [304]
3. Divergence	$J = \int_{-\infty}^{\infty} \ln(\frac{f(x)}{g(x)}) (f(x) - g(x)) dx$	No	No	Yes	[181] [182] [188]
4. Bhattacharyya	$B = -\ln \int_{-\infty}^{\infty} (f(x) g(x))^{1/2} dx$	No	No	Yes	[188] [29]
5. Jeffreys-Matusita	$M = \{\int_{-\infty}^{\infty} (\sqrt{g(x)} - \sqrt{f(x)})^2 dx\}^{1/2}$	Yes	Yes	Yes	[181] [182] [259]
6. Kolmogorov-Variational Distance	$K(p) = \int_{-\infty}^{\infty} \|p_g g(x) - p_f f(x)\| dx$	Yes	Yes	Yes	[188] [3] [210]
7. Kullback-Liebler Numbers	$L_{fg} = \int_{-\infty}^{\infty} \ln(\frac{f(x)}{g(x)}) f(x) dx$	No	No	Yes	[216] [188]

TABLE 1.3 (con't.)

Name	Form	Metric in			
		C	MVN	MVN$_\Sigma$	
8. Swain-Fu	$T = \dfrac{\lvert\mu_f - \mu_g\rvert}{D_f + D_g}$ where $D_f = \left\{ \dfrac{\lvert\mu_f-\mu_g\rvert^2\,(p+2)}{tr[\Sigma_f(\mu_f-\mu_g)(\mu_f-\mu_g)^T]} \right\}^{1/2}$	No	No	Yes	[353]
9. Mahalanobis	$\Delta = \{(\mu_g-\mu_f)^T \Sigma^{-1}(\mu_g-\mu_f)\}^{1/2}$	*	*	Yes	[251] [252]
10. Samuels-Bachi	$\mu = \{\int_0^1 [F^{-1}(\alpha) - G^{-1}(\alpha)]d\alpha\}^{1/2}$ where $F^{-1}(\alpha) = \inf\{c \mid Q_c \cap Q_\alpha \neq \emptyset\}$ and $Q_c = \{x \mid \sum_{i=1}^p x_i \le c\}_j, Q_\alpha = \{x \mid F(x) \ge \alpha\}$	No	No	No	[307]
11. Kiefer-Wolfowitz	$V = \int_{-\infty}^{\infty} \lvert F(x) - G(x)\rvert\, e^{-\lvert x\rvert}dx$ where $\lvert x\rvert$ = vector norm of x.	Yes	Yes	Yes	[197]

*Not defined in these cases.

1.6. Clustering Methods Based on Euclidean Distance

A substantial amount of effort in the development of clustering and classification methods has been directed towards the construction of methods that rely on the minimization of the within-group (error) sum of squares. These methods have been termed minimum-variance constraint methods [396] and can be described and illustrated in terms of the squared Euclidean distance. This section will contain a discussion of various clustering techniques which can be put into this framework. Futhermore, it will be seen that many such clustering techniques can be utilized into the same clustering algorithm by means of a general relationship involving the distance measures d_{ij}.

Consider the data matrix $X = (X_1, X_2, \ldots, X_n)$. The squared Euclidean distance between X_i and X_j is given by

$$d_{ij}^2 = (X_i - X_j)^T (X_i - X_j).$$

Various clustering methods which are actually based on this measure of distance will now be described. The descriptions are rather brief and for greater detail the reader is referred to the indicated references.

Sorensen [338] describes a method which involves clustering by complete linkage. By this is meant that no two individuals in a group (cluster) have a similarity which is less than a threshold value S. In terms of the squared Euclidean distance d^2 this means that the distance between any two cluster points (individuals) must not exceed a threshold value r. The threshold distance r defines the maximum permitted diameter of the cluster subset.

MacNaughton-Smith [234] defines a heirarchical method, called the furthest neighbor technique, which is similar to Sorensen's. This method treats each individual as a single-point cluster. The

objects are combined in steps such that at each step the two clusters
having the smallest Euclidean distance between the two most distant
objects, one from each cluster, are combined. The procedure involves
n - 1 steps and obtains all possible groupings which can be obtained
by Sorensen's method for any threshold value.

Ward [387] uses the within-group sum of squares as the objective
function. His criterion is also termed the error sum of squares
(ESS) and is just the sum of squares of the distances from each
point to its parent cluster mean. His method is also a hierarchical
process which combines at each step those two clusters which result
in the least increase in the WGSS objective function. The increase
in WGSS in combining clusters I $(n_1$ objects) and $J(n_2$ objects) to
form one cluster is given by

$$D_{IJ} = \frac{n_1 n_2}{n_1 + n_2} \ (\bar{X} - \bar{Y})^T (\bar{X} - \bar{Y})$$

according to section 1.5, where $\bar{X}$ and $\bar{Y}$ denote the mean vectors
corresponding to clusters I and J. Ward's method favors the grouping
of small close clusters.

Sokal and Michener [334] describe a procedure, called the
centroid method, for which the distance between two clusters I and J
is given in terms of the squared Euclidean distance between their
centroids, that is, $d_{IJ}^2 = (\bar{X} - \bar{Y})^T (\bar{X} - \bar{Y})$. This method can be
described as being a heirarchical procedure [223] such that the
two clusters I and J having the minimum d_{IJ}^2 are combined at each of
n - 1 steps. If n_1 is much larger than n_2 then the centroids of
I ∪ J and I are about the same and the characteristics of J are lost.
This prompted the term "weighted group" technique.

Another technique due to Sokal and Michener [334], called the
pair-group technique uses the relationship between a single object

i and a cluster I. This relationship is defined to be the average of the similarities between the object i and all the objects in cluster I. To relate this average similarity to Euclidean distance let the vectors corresponding to the objects in I be denoted by $X_1, X_2, \ldots, X_{n_I}$ and let $\bar{X}$ denote the centroid of I. Then the average distance from the object $i \notin I$ to all the objects in I, denoted by D_{iI} is

$$D_{iI} = \frac{1}{n_I} \sum_{j=1}^{n_I} (X_j - Y)^T (X_j - Y)$$

where Y denotes the vector corresponding to $i \notin I$. Thus

$$D_{iI} = \frac{1}{n_I} \sum_{j=1}^{n_I} (X_j - \bar{X} + \bar{X} - Y)^T (X_j - \bar{X} + \bar{X} - Y)$$

$$= \frac{1}{n_I} \sum_{j=1}^{n_I} (X_j - \bar{X})^T (X_j - \bar{X}) + (\bar{X} - Y)^T (\bar{X} - Y) \qquad (1.11)$$

The first term on the right hand side, denoted by s_I^2, is the <u>within variance</u> for the objects in I and the second term is the distance squared from the object i to the centroid of I. In a sequential manner then, that object $i \notin I$ for which D_{iI} is a minimum is fused with the cluster I. From (1.11) one can see that if two clusters have comparable variance then the average distance D_{iI} minimizes the distance from an object i to a cluster centroid I. For clusters having different variances the measure (1.11) favors fusion with one of the smaller variance clusters.

Lance and Williams [225] extend the concept of the pair-group technique and define the average similarity between two clusters I and J to be the average of the similarities between all pairs of objects, one from each cluster. Their technique is termed the <u>group-average</u> technique. Clusters are formed heirarchically, two clusters

I and J being combined when the average is a minimum. To relate average similarity to Euclidean distance let X and Y denote the means for the clusters I and J, respectively. The average squared distance between objects in I and objects in J, denoted by D_{IJ}^2, is

$$D_{IJ}^2 = \frac{1}{n_I n_J} \sum_{i=1}^{n_I} \sum_{j=1}^{n_J} (X_i - Y_j)^T (X_i - Y_j)$$

$$= \frac{1}{n_I} \sum_{i=1}^{n_I} D_{iJ}$$

$$= \frac{1}{n_I} \sum_{i=1}^{n_I} \{ \frac{1}{n_J} \sum_{j=1}^{n_J} (Y_j - \bar{Y})^T (Y_j - \bar{Y}) + (\bar{Y} - X_i)^T (\bar{Y} - X_i) \}$$

$$= \frac{1}{n_J} \sum_{j=1}^{n_J} (Y_j - \bar{Y})^T (Y_j - \bar{Y}) + \frac{1}{n_I} \sum_{i=1}^{n_I} (\bar{Y} - X_i)^T (\bar{Y} - X_i)$$

The first term on the right hand side is the within variance for cluster J and the second term is the average squared distance from $\bar{Y}$ to X_i, $i=1,2,\ldots,n_I$. Thus the second term can be written as

$$s_I^2 + d^2(\bar{X},\bar{Y}) = \frac{1}{n_I} \sum_{i=1}^{n_I} (X_i - \bar{X})^T (X_i - \bar{X}) + (\bar{Y} - \bar{X})^T (\bar{Y} - \bar{X}).$$

Thus

$$D_{IJ}^2 = s_I^2 + s_J^2 + d^2(\bar{X},\bar{Y}), \tag{1.12}$$

and the minimization of the average similarity is equivalent to the minimization of (1.12).

Bonner [33] describes a method for which an object is chosen at random and used as a starting point. All those objects which are within r units from the starting point are fused into the first cluster. From the remaining points another is chosen at random and the process is repeated as before. After all the points have been allocated, each is assigned to its nearest cluster center to form disjoint groups.

Hyvarinen [171] defines a process identical to Bonner's, except that "typical" points are chosen to initialize clusters, rather than choosing points at random. He uses an information-loss statistic to detect the "typical" points, however, in terms of d^2, that point nearest the centroid of the residual set might suffice.

A procedure of Ball and Hall [18] forms k clusters by first selecting k objects at random and then assigning each of the remaining n - k objects to that center which is nearest it. The cluster centroids are computed and any two clusters I and J are combined if D_{IJ}^2 is less than a threshold value r. A cluster is split if the variance S_x^2 within the cluster of any one variable x exceeds a threshold S^2. The variance S_I^2 of each resultant cluster I is thus constrained by

$$S_I^2 \leq pS^2$$

where p is the number of variables. The cluster centroids replace the original cluster centers and the process is continued until convergence is attained. The Ball and Hall procedure has come to be quite popular.

MacQueen [237] proposed a technique similar to Ball and Hall's. His procedure chooses k objects at random to be used as cluster centers. Each object is assigned to the center nearest it if its distance from that cluster center does not exceed a threshold value r. If an object is farther than r units from all centers then it initialize a new cluster. After each allocation, the cluster centroid is recomputed and becomes the new cluster center. When the distance between two centroids becomes less than another threshold value, then the two corresponding clusters are combined. The process continues until convergence is attained.

A method due to Sebestyen [315] is similar to that of MacQueen. However, Sebestyen's procedure allocates a point to a cluster center if the point is less than r units from the center, but if the distance exceeds R units (R > r) then the point initializes a new cluster center. If the distance d is such that r < d ⩽ R, then the point is set aside for allocation at a later iteration.

Jancey [174] proposes a technique which is similar to MacQueen's. Jancey's procedure selects k points for centers from E_p at random, rather than k random objects from the n. The objective function which is minimized is the within group sum of squares.

Forgey [114] proposes a method similar to Jancey's which yields final groupings similar to those of Jancey's method. This method also minimizes the within group sum of squares.

The main reasons for the popularity of the Euclidean metric are probably that it is very appealing intuitively and directly related to WGSS as indicated in equation (1.7).

There are various objections to the minimum-variance approach to cluster analysis. For example, changes in scale will modify the resultant clusters. For a discussion of this and other objections see Wishart [396]. His paper also contains a discussion of the methods outlined above. Friedman and Rubin [122] discuss some invariant criteria for grouping data.

1.7. An Algorithm for Heirarchical Clustering

The scheme in hierarchical clustering is to consider $I = \{I_1, I_2, \ldots, I_n\}$ as a set of clusters $\{I_1\}$, $\{I_2\}, \ldots, \{I_n\}$ and then select those two clusters, say I_i and I_j, which are nearest in some sense, and "clumping" or "fusing" them into one cluster. The new cluster set of n-1 clusters becomes

$$\{I_1\}, \{I_2\},\ldots, \{I_i,I_j\} \ldots \{I_n\}.$$

By repeating this process, we can produce sequentially cluster sets of n-2 clusters, n-3 clusters, and so on, so that finally we would have a cluster set of n individuals which will be the original set $I = \{I_1,I_2,\ldots,I_n\}$.

For our purposes here we will consider the squared Euclidean distance d_{ij}^2 as our measure of distance. For display purposes one can compute the matrix $D = \{d_{ij}^2\}$ where d_{ij}^2 is the squared distance between I_i and I_j.

TABLE 1.4

The Values of d_{ij}^2

	I_1	I_2	I_3 $\cdots$ I_n
I_1	0	d_{12}^2	d_{13}^2 $\cdots$ d_{1n}^2
I_2		0	d_{21}^2 $\cdots$ d_{2n}^2
I_3			0 $\cdots$ d_{3n}^2
I_n			0

Suppose that I_i and I_j are nearest, that is, $d_{ij}^2 = \min\{d_{ij}^2, i \neq j\}$, so that I_i and I_j are fused into a new cluster $\{I_i,I_j\}$. A new $(n-1)\times(n-1)$ matrix can now be computed and displayed as in Table 1.5.

TABLE 1.5

The Values of d_{ij}^2 After a Single Fusion

	$\{I_i,I_j\}$	I_1	I_2	I_3 $\cdots$ I_n
$\{I_i,I_j\}$	0	d_{ij1}^2	d_{ij2}^2	d_{ij3}^2 $\cdots$ d_{ijn}^2
I_1		0	d_{12}^2	d_{13}^2 $\quad$ d_{1n}^2
I_2			0	d_{23}^2 $\quad$ d_{2n}^2
I_3 $\vdots$ I_n				0 $\quad$ d_{3n}^2 $\ddots$ d_{nn}^2

One notes that n-2 rows of the new data matrix can be taken directly without recalculation from the old data matrix, while one row, the first row, must be computed anew. However, if one could write a formula for d_{ijk}^2, k = 1,2,...,n, k ≠ i ≠ j as a function of the elements of the data in the previous matrix then the amount of computation could be perhaps minimized.

Lance and Williams [223] have given a recursive scheme so that the calculation of the distance matrix depends only on the elements of the distance matrix at the previous step. Their recursive scheme includes nearest neighbor, furthest neighbor, median, group average, and centroid philosophies of measuring distance. All five schemes except nearest neighbor and median have been described in section 1.6. The nearest neighbor distance between two clusters I and J has been defined in section 1.5 in which case, the scheme is to group those two clusters for which the distance between them is a minimum. The median philosophy is the same as the centroid philosophy except that in combining two clusters I and J it is assumed that $n_I = n_J$ so that the centroid of the resultant cluster lies midway between the centroids of I and J.

Wishart [394] demonstrated that the procedure of Ward [387] can be incorporated into the same scheme as the five procedures mentioned above. From section 1.5 we observe that the increase in WGSS in combining clusters I and J, denoted by W_{IJ}, is given by

$$W_{IJ} = \frac{n_I n_J}{n_I + n_J} (\bar{X}_I - \bar{X}_J)^T (\bar{X}_I - \bar{X}_J)$$

$$= \frac{n_I n_J}{n_I + n_J} d_{IJ}^2 \qquad (1.13)$$

where $d_{IJ}^2 = (\bar{X}_I - \bar{X}_J)^T (\bar{X}_I - \bar{X}_J)$. If $I \cup J = L$ is combined with K then it can be shown that $d_{KL}^2 = (\bar{X}_K - \bar{X}_L)^T (\bar{X}_K - \bar{X}_L)$ is given by

$$d_{KL}^2 = \frac{n_I}{n_L} d_{KI}^2 + \frac{n_J}{n_L} d_{KJ}^2 - \frac{n_I n_J}{n_L^2} d_{IJ}^2 \qquad (1.14)$$

Furthermore, from (1.13) we have

$$W_{KL} = \frac{n_K n_L}{n_K + n_L} d_{KL}^2 . \qquad (1.15)$$

Thus on substituting for each d^2 from (1.15) into (1.14) we obtain

$$W_{KL} = \frac{1}{n_K + n_L} \{ (n_I + n_K) W_{KI} + (n_J + n_K) W_{KJ} - n_K W_{IJ} \} \qquad (1.16)$$

Equation (1.16) gives the increase in WGSS when K is combined with $I \cup J$.

Starting with the matrix of squared Euclidean distances (Table 1.4), $D = d_{ij}^2$, $i = 1,2,\ldots,n$; $j = 1,2,\ldots,n$, the procedure is to combine those two objects I_p and I_q for which $d_{pq}^2 = 2W_{pq}$ is a minimum. The single-object cluster I_p is then replaced by $I_p \cup I_q$ and the distances d_{ip}^2, $i = 1,2,\ldots,n$; $i \neq p,q$, of the matrix D are replaced by $d_{ip}^2 = 2W_{ip}$. All the elements in the qth column and row of D are set equal to zero, i.e. S_q is "inactive" [394]. Also, n_p is replaced by $n_p + n_q$ and n_q is set equal to zero. The equation

$$d_{ip}^2 = 2W_{ip} \qquad\qquad (1.17)$$

holds for all $\{d_{ij}^2\}$, $i,j \neq q$.

By substituting W_{ip} (actually W_{ir}) from equation (1.16) into equation (1.17), we have

$$d_{ip}^2 \leftarrow 2W_{ir} = \frac{2}{(n_i+n_r)} \ [(n_i+n_p)W_{ip} + (n_i+n_q)W_{iq} - n_i W_{pq}]$$

$$= \frac{1}{(n_i+n_r)} \ [(n_i+n_p)d_{ip}^2 + (n_i+n_q)d_{iq}^2 - n_i d_{pq}^2], \quad (1.18)$$

where $n_r = n_p + n_q$. If, at every fusion step the elements of the pth column and row of D are modified according to equation (1.18), then equation (1.17) will hold for all d_{ij}^2 for active sets S_i and S_j. Note that the quantity d_{ij}^2 in (1.17) is not a Euclidean distance unless one is dealing with two single-object clusters.

The fusion algorithm may be summarized as follows [394]:

(i) Find $d_{pq}^2 = \min_{i,j} \{d_{ij}^2\}$, $i = 1,\ldots,j-1$; $j = 2,\ldots,n$;

$n_i > 0$; $n_j > 0$.

(ii) The increase in objective function caused by the fusion of I_p and I_q is $W_{pq} = \frac{1}{2} d_{pq}^2$.

(iii) I_p is replaced with $S_p \cup S_q$ by modifying the row $\{d_{ip}^2\}$, $i = 1,2,\ldots,p-1$; $n_i > 0$ and column $\{d_{pj}^2\}$, $j = p + 1,\ldots,n$; $j \neq q$; $n_j > 0$ of D according to equation (1.18).

(iv) Set $n_p \equiv n_p + n_q$ and $n_q = 0$ to render S_q inactive.

(v)   Reclassify the elements of cluster S_q into cluster S_p, return to (i) and iterate for n - 2 steps.

Lance and Williams [223] have derived a relationship of the type given by equation (1.18) for five other common clustering processes mentioned above.   The general form of the relationship in equation (1.18) is given by

$$d_{hk}^{2} = \alpha_i \, d_{hi}^{2} + \alpha_j \, d_{hj}^{2} + \beta \, d_{ij}^{2} + \gamma |d_{hi}^{2} - d_{hj}^{2}| \qquad (1.19)$$

where α_i, α_j, β, and γ are determined by the particular clustering process.   In equation (1.19) d_{hk} denotes the measure of distance between I_h and $I_K = I_i \cup I_j$.

The values of the parameters in the general form (1.19) are summarized below for six different clustering processes:

Nearest Neighbor: $\alpha_i = \alpha_j = \frac{1}{2}$; $\beta = 0$; $\gamma = -\frac{1}{2}$.

Furthest Neighbor: $\alpha_i = \alpha_j = \frac{1}{2}$; $\beta = 0$; $\gamma = \frac{1}{2}$.

Median: $\alpha_i = \alpha_j = \frac{1}{2}$; $\beta = -\frac{1}{4}$; $\gamma = 0$.

Group Average: $\alpha_i = n_i/n_k$; $\alpha_j = n_j/n_k$; $\beta = \gamma = 0$.

Centroid: $\alpha_i = n_i/n_k$; $\alpha_j = n_j/n_k$; $\beta = -\alpha_i\alpha_j$; $\gamma = 0$.

Ward's Method: $\alpha_i = \frac{n_h+n_i}{n_h+n_k}$; $\alpha_j = \frac{n_h+n_j}{n_h+n_k}$; $\beta = \frac{-n_h}{n_h+n_k}$; $\gamma = 0$.

The first five results may be found in Lance and Williams [223].   The result for Ward's method was obtained by Wishart [394] and follows from equation (1.19).   All six methods have been incorporated into the same computer program ([398],[399]) by varying the parameters α, β, and γ.

1.8.   Other Aspects of the Cluster Problem

An important item in solving the cluster problem is determining the number of clusters that are desired in the solution. In some cases it may be possible to choose the number of clusters, m, a priori, however, in the more general problem the number of clusters are determined by the clustering process. This monograph will not be concerned with the very difficult problem of determining the number of clusters.

It is well-known that some large data problems yield to random techniques for practical solutions. Fortier and Solomon [119] considered such techniques and found that laws of simple random sampling can be used to compute how many cluster selections are to be attempted and enumerated before one attains a probability α of having captured one of the cluster sets which would rank among the best. The number of selections desired is thus a function of a pre-determined β, the proportion of all the selections which can be labled "best" or acceptable in some sense. In the setting considered here the total scatter of cluster sets larger than some preassigned constant determine those cluster sets which are considered good and these together are a fraction β of the total number of possible clusterings. Fortier and Solomon calculated a table of $S(\alpha,\beta)$, the number of samplings required to get a good clustering not for scatter but for a "measure of belonging" defined by Holzinger and Harman [168] [168], and this table is reproduced here. They concluded that simple random sampling is in general not an effective way when the distribution of the index is very skewed and favorable values of the index are in the tail, but do suggest "modifications of the sampling strategy may improve the situation and these should be studied".

TABLE 1.6

Table of $S(\alpha,\beta)$ for values of α and β.

β \ α	.20	.10	.05	.01	.001	.0001
.20	8	11	14	21	31	42
.10	16	22	29	44	66	88
.05	32	45	59	90	135	180
.01	161	230	299	459	689	918
.001	1,626	2,326	3,026	4,652	6,977	9,303
.0001	17,475	25,000	32,526	50,000	75,000	100,000

Two tacit assumptions underlie the problem in cluster analysis; (1) the characteristics selected are those which have relevance to the desired cluster solution, and (2) the units of measure (scale on which we take our measurements) are "valid". The first problem is what has been called the characteristic or feature selection problem for which there exists some studies [229], [230], and [255]. Generally, the decision as to what characteristics of the individual one should measure has been assumed resolved prior to the clustering process. However one should be aware that this arbitrariness always exists and in some cases should be considered.

The problem of scaling is always with us and indeed is bothersome. The general solution is to normalize the data by centering the data and then dividing by a scale factor so that the variance is one. Problems of interpretation always present themselves when one strays from original (popular) units. However, more serious is that the clusters will vary under different scales. What is desirable is a cluster technique which remains invariant under various families of scaling transformations.

Chapter 2

CLUSTERING BY COMPLETE ENUMERATION

2.1. Introduction

A direct way of solving the cluster problem is to evaluate the objective function for each choice of clustering alternatives and then choose the partition yielding the optimal (minimum) value of the objective function. However, this procedure is not practical and is virtually impossible unless n (the number of objects) and m (the number of clusters) are small. This procedure is termed clustering by complete enumeration. If n = 8 and m = 4, for example, then the number of clustering alternatives is 1701; that is, there are 1701 ways of partitioning 8 objects into 4 subsets. The number of clustering alternatives, denoted by $S(n,m)$, is a Stirling's number of the second kind and can be computed by means of a formula to be derived in the next section. This chapter is only incidently related to the cluster problem and may be omitted by the reader.

2.2. The Number of Partitions of n objects into m Nonempty Subsets

The process of partitioning a set of n objects into m nonempty subsets can be viewed as that of placing n distinct balls into m identical boxes with no box empty. Let w denote the total number of such partitions. Then w m! is the total number of ways of placing n distinct objects into m distinct boxes with no box empty.

One of the most efficient methods of solving problems related to the distribution of n balls among m boxes (or partitioning a set of n objects into m nonempty subsets) is the method of generating functions.

Consider first the function

$$(1 + x_1 t)(1 + x_2 t)\ldots(1 + x_n t). \qquad (2.1)$$

The expansion of (2.1) yields a polynomial in t such that the coefficient of t^k yields all the combinations of n objects taken k at a time. If $x_1 = x_2 = \ldots = x_n = 1$ then the coefficient of t^k yields the quantity $\binom{n}{k}$, that is, the number of ways of choosing k items from the n items. The sequence $\binom{n}{k}$, k = 0,1,...,n can thus be generated by the <u>generating function</u>,

$$(1 + t)^n = \sum_{k=0}^{n} \binom{n}{k} t^k.$$

If one is interested in the order of objects in each combination then the function

$$(1 + t)^n = \sum_{k=0}^{n} \binom{n}{k} t^k = \sum_{k=0}^{n} {}_nP_k \frac{t^k}{k!} \qquad (2.2)$$

can be used to generate the sequence ${}_nP_k = k! \binom{n}{k}$, k = 0,1,...,n. In this case the generating function (2.2) is called an <u>exponential generating function</u>. If repetitions are allowed then the enumerator for any object is a series containing a term $t^k/k!$ for each k in the specification of repetitions. Thus if unlimited repetition is allowed, then the enumerator for each object is given by

$$1 + t + \frac{t^2}{2!} + \ldots = e^t.$$

Consequently the number of permutations of n distinct objects taken k at a time is given by the coefficient of $t^k/k!$ in the generating function

$$\left(1 + t + \frac{t^2}{2!} + \ldots\right)^n = e^{nt} = \sum_{k=0}^{\infty} n^k \frac{t^k}{k!} \quad .$$

The required number of permutations is thus n^k. If unlimited

repetition is allowed with each object occurring at least once then
the enumerator is given by

$$(t + \frac{t^2}{2!} + \ldots)^n = (e^t - 1)^n$$

$$= \sum_{j=0}^{n} \binom{n}{j} (-1)^j e^{(n-j)t}$$

$$= \sum_{k=0}^{\infty} \frac{t^k}{k!} (\sum_{j=0}^{n} \binom{n}{j} (-1)^j (n-j)^k). \qquad (2.3)$$

The number of permutations in this case is given by

$$\sum_{j=0}^{n} \binom{n}{j} (-1)^j (n-j)^k . \qquad (2.4)$$

The generating function used to enumerate the total number of
ways of distributing n distinct balls into m distinct boxes when
the order of the balls in each box is immaterial and such that box
i contains n_i balls, $i = 1,2,\ldots,m$, is given by

$$(x_1 + x_2 + \ldots + x_m)^n = \sum \frac{n!}{n_1! \; n_2! \; \ldots \; n_m!} \; x_1^{n_1} \; x_2^{n_2} \; \ldots \; x_m^{n_m} ,$$

$$(2.5)$$

where the summation is extended over all ordered $-m$ partitions of
n, that is $n_1 + n_2 + \ldots + n_m = n$. The coefficient of $x_1^{n_1} \; x_2^{n_2} \ldots$
$x_m^{n_m}$ yields the distribution such that there are n_1 balls in box 1,
n_2 balls in box 2,$\ldots,n_m$ balls in box m. Thus

$$\frac{n!}{n_1! \; n_2! \; \ldots \; n_m!} = \binom{n}{n_1 \; n_2 \; \ldots \; n_m} \qquad (2.6)$$

yields the number of ways of placing n_i balls in box i, $i = 1,2,\ldots,m$.

The number of ways of placing n distinct balls into m distinct
boxes can be arrived at by use of the generating function

$$\prod_{i=1}^{m} (1 + x_i t + x_i^2 \frac{t^2}{2!} + \ldots + x_i^n \frac{t^n}{n!}) . \qquad (2.7)$$

In this case the ith factor

$$1 + x_i t + x_i^2 \frac{t^2}{2!} + \ldots + x_i^n \frac{t^n}{n!} \qquad (2.8)$$

corresponds to the <u>enumerator for the</u> i^{th} <u>box</u>. The coefficient of t^n in the expansion of (2.7) is given by

$$\sum \frac{1}{n_1! \, n_2! \, \ldots \, n_m!} x_1^{n_1} x_2^{n_2} \ldots x_m^{n_m}$$

where the summation extends over all ordered $-m$ partitions of n. The coefficient of $t^n/n!$ in (2.7) is thus given by

$$\sum \frac{n!}{n_1! \, n_2! \, \ldots \, n_m!} x_1^{n_1} x_2^{n_2} \ldots x_m^{n_m} = (x_1 + x_2 + \ldots + x_m)^n .$$

Now, by (2.6) $\dfrac{n!}{n_1! \, n_2! \, \ldots \, n_m!}$ is the number of ways of distributing

n balls in m boxes so that box k contains n_k balls ($k = 1,2,\ldots,m$) while $x_1^{n_1} x_2^{n_2} \ldots x_m^{n_m}$ describes the particular distribution. By setting $x_1 = x_2 = \ldots = x_m = 1$ in (2.7) it follows that the coefficient of $t^n/n!$ in

$$(1 + t + t^2/2! + \ldots + t^n/n!)^m$$

is the number of ways of placing n distinct balls into m distinct boxes. However, this is the same as the coefficient of $t^n/n!$ in

$$(1 + t + t^2/2! + \ldots)^m = e^{mt} .$$

Thus, the number of ways of placing n distinct balls into m distinct boxes is m^n.

The number of partitions of a set of n objects into m subsets, none of which is empty, is given by the following theorem.

__Theorem 1.1.__ The number of ways of putting n distinct balls into m distinct boxes such that none of the m boxes is empty is given by

$$\sum_{j=0}^{m} \binom{m}{j} (-1)^j (m-j)^n,$$

when the order of the balls within each box is irrelevant.

__Proof__: The enumerator for occupancy of the ith box is given by

$$(x_i t + x_i^2 t^2/2! + \ldots) = e^{tx_i} - 1,$$

since each box must contain at least one ball. The required exponential generating function, denoted by $E(t, x_1, \ldots, x_m)$, is given by

$$E(t, x_1 x_2, \ldots, x_m) = \prod_{i=1}^{m} (e^{tx_i} - 1). \tag{2.9}$$

The number of ways of placing n balls in m boxes so that no box is empty is given by the coefficient of $t^n/n!$ when $x_1 = x_2 = \ldots = x_m = 1$. The generating function (2.9) becomes

$$E(t, 1, 1, \ldots, 1) = (e^t - 1)^m, \tag{2.10}$$

which is the same form as (2.3). Thus

$$E(t, 1, 1, \ldots, 1) = \sum_{k=0}^{\infty} \frac{t^k}{k!} [\sum_{j=0}^{m} \binom{m}{j} (-1)^j (m-j)^k].$$

The coefficient of $t^n/n!$ is given by

$$\sum_{j=0}^{m} \binom{m}{j} (-1)^j (m-j)^n$$

which yields the required result.

In Theorem 1.1 the m boxes are assumed to be distinct. However, in the partitioning of the n objects into m subsets, none of which is empty, the order of the m subsets is irrelevant. Consequently, from this fact and Theorem 1.1 it follows that the total number of ways of partitioning n objects into m subsets is given by

$$w = \frac{1}{m!} \sum_{j=0}^{m} \binom{m}{j} (-1)^j (m-j)^n. \tag{2.11}$$

It remains to show the relationship between the result (2.11) and Stirling's numbers of the second kind.

Stirling's numbers of the second kind arise in the calculus of finite differences and provide a way of expressing powers of x in terms of "falling factorials". The falling factorial is defined by

$$x_{(0)} = 1, \quad x_{(n)} = x(x-1) \ldots (x-n+1), \quad n = 1, 2, \ldots \; .$$

Definition 2.1. Stirling's numbers of the second kind are defined to be those numbers $S(n,i)$ such that

$$x^n = \sum_{i=1}^{n} S(n,i) x_{(i)}, \quad S(n,0) = 0,$$

and

$$S(n, n+k) = 0 \text{ for } k > 0.$$

Now, Newton's Theorem ([172], chapter 6) states that if $U(x)$ is an n-th degree polynomial in x, then it may be written in the form

$$U(x) = \sum_{k=0}^{n} x_{(k)} \ [\frac{\Delta^k U(x)}{k!}]_{x=0} \ , \tag{2.12}$$

where Δ is defined by

$$\Delta f(x) = f(x+1) - f(x),$$

$$\Delta^k f(x) = \Delta [\Delta^{k-1} f(x)],$$

and

$$\Delta^0 f(x) = f(x).$$

Expanding $U(x) = x^n$ according to (2.12) we obtain

$$x^n = \sum_{m=0}^{n} x_{(m)} \ [\frac{\Delta^m x^n}{m!}]_{x=0} \ .$$

From the definition of $S(n,i)$ we then have

$$S(n,m) = [\frac{\Delta^m x^n}{m!}]_{x=0}$$

$$= \frac{1}{m!} [\Delta^m x^n]_{x=0} \ .$$

It remains to show that

$$\sum_{j=0}^{m} \binom{m}{j} (-1)^j (m-j)^n = [\Delta^m x^n]_{x=0} \ .$$

To show this we make use of a shift operator $E = 1 + \Delta$, with $Ef(x) = f(x+1)$, $E^j f(x) = f(x + j)$, and $\Delta = E - 1$. From this we obtain

$$\sum_{k=0}^{m} \binom{m}{k} (-1)^k (m-k)^n = \sum_{k=0}^{m} \binom{m}{k} (-1)^k E^{m-k} x^n \Big|_{x=0}$$

$$= (E - 1)^m x^n \Big|_{x=0}$$

$$= [\Delta^m x^n]_{x=0}$$

Thus we have $w = S(n,m)$.

If the number of subsets m is known in advance the total number of partitions or clustering alternatives is given by $S(n,m)$. However, if m is not specified then the total number of clustering alternatives is given by

$$\sum_{m=1}^{n} S(n,m) \ .$$

2.3. Recursive Relation for Stirling's Numbers of the Second Kind

Equation (2.11) may be used to evaluate Stirling's numbers of the second kind. However, if a table of such values is desired then it may be more feasible to use the recurrence relation

$$S(n+1,i) = i \ S(n,i) + S(n, i - 1) \ .$$

This recurrence relation is derived by first observing that

$$x_{(i+1)} = x(x-1) \ \cdots \ (x-i)$$

$$= x_{(i)} (x-i)$$

$$= x \ x_{(i)} - i \ x_{(i)} \ .$$

Thus

$$x_{(i+1)} + i \ x_{(i)} = x \ x_{(i)} \ . \tag{2.13}$$

Using the definition of S(n,i) we have

$$x^{n+1} = \sum_{i=1}^{n+1} S(n+1,i)x_{(i)} \tag{2.14}$$

and

$$x^{n+1} = x \, x^n = \sum_{i=1}^{n} S(n,i)x \, x_{(i)} \ . \tag{2.15}$$

Using identity (2.13) in (2.15) we obtain

$$x^{n+1} = \sum_{i=1}^{n} S(n,i)(x_{(i+1)} + i \, x_{(i)})$$

$$= \sum_{i=2}^{n+1} S(n, \, i-1)x_{(i)} + \sum_{i=1}^{n} i \, S(n,i)x_{(i)} \ .$$

But since $S(n,0) = S(n,n+1) = 0$ we may write

$$x^{n+1} = \sum_{i=1}^{n+1} [S(n,i-1) + i \, S(n,i)]x_{(i)} \tag{2.16}$$

and comparing this result with (2.14) we have

$$S(n+1,i) = S(n,i-1) + i \, S(n,i).$$

An equivalent form for S(n,m) is given by

$$S(n,m) = \frac{1}{m!} \sum_{k=0}^{m} \binom{m}{k} (-1)^{m-k} k^n$$

from which we obtain

$$\frac{S(n,m)}{m^n} = \frac{1}{m!} \sum_{k=0}^{m} \binom{m}{k} (-1)^{m-k} \left(\frac{k}{m}\right)^n \ .$$

As n → ∞ every term in the sum vanishes except the last so that

$$\lim_{n \to \infty} \frac{S(n,m)}{m^n} = \frac{1}{m!} \quad .$$

Asymptotically, we then obtain

$$S(n,m) \sim \frac{m^n}{m!} \quad .$$

Table 2.1 gives S(n,m) for values of n up to 8.

TABLE 2.1

Number of Clustering Alternatives for Various Values of n^1 and m^1

n \ m	1	2	3	4	5	6	7	8
1	1							
2	1	1						
3	1	3	1					
4	1	7	6	1				
5	1	15	25	10	1			
6	1	31	90	65	15	1		
7	1	63	301	350	140	21	1	
8	1	127	966	1701	1050	266	28	1

[1] n and m represent the number of objects and clusters, respectively.

2.4. Computational Aspects of Complete Enumeration

Given an objective function, such as within group sum of
squares or, equivalently, Euclidean distance, the optimal solution
to the cluster problem can be obtained, conceptually at least, by
evaluating the objective function for each clustering alternative

and then choosing that alternative which yields the optimal value of the objective function. This process is impractical unless n is small. It is obvious that any two clustering alternatives can contain various subsets which are alike so that there may be extensive computations which are duplicated under complete enumeration.

Thus, in clustering under complete enumeration it is desirable to develop some scheme or algorithm that eliminates redundant or unnecessary computations. This factor alone suggests the need for a dynamic programming algorithm that will eliminate the redundant computations of complete enumeration and at the same time converge on the optimal solution. Various existing clustering techniques yield rapid solutions to the cluster problem. Hierarchical techniques are of this type. However, these techniques operate on subclasses of clustering alternatives and there is no guarantee that the solution is the optimal one or even close to the optimal one.

In clustering by complete enumeration it is possible to store the data matrix X and operate directly on it without having to resort to auxilary storage. However, the amount of computation necessary warrants the task virtually hopeless in spite of the high speed computers presently available. (Table 2.1 gives the number of clusters for n (the number of objects) up to 8.)

Attempts to develop dynamic programming techniques to solve the cluster problem will probably require rapid access storage. A dynamic programming technique will have to keep track of "old" computations as well as perform "new" ones thus making rapid access storage highly desirable if not necessary. One such dynamic programming technique has been developed by Jensen [183] and is described in the next chapter. Other mathematical programming based techniques to perform cluster analysis are also discussed.

Chapter 3

MATHEMATICAL PROGRAMMING AND CLUSTER ANALYSIS

Recall that the objective in solving the cluster problem is to determine the optimal partitioning of n objects into m disjoint subsets in such a manner that a certain criterion of homogeneity within clusters is satisfied. A way of accomplishing this objective as discussed in Chapter 2 is by complete enumeration, i.e., examine the homogeneity criterion for all possible partitions into m clusters and choose that one which is optimal. Unfortunately, the method of complete enumeration is in general impractical, even for small values of n and m.

An alternative to the complete enumeration technique is to ultilize some of the techniques properly called mathematical programming techniques in an attempt to reduce the amount of computation but yet converge hopefully to an optimal solution. Most of the techniques previously mentioned search for the optimal solution in a small class of subsets (clusters) and there is no quarantee that the solution is optimal over the whole class of clusters. Various applications of mathematical programming have appeared in the literature, e.g. Jensen [183], Vinod [380], and Rao [291].

3.1 Application of Dynamic Programming to the Cluster Problem

In this section we consider the problem of partitioning a set of 6 objects into 3 subsets when the distance between two objects is the Euclidean metric or the criterion is the minimization of Within Groups Sum of Squares (WGSS).

Recall that WGSS is given by

$$W = \text{tr} \sum_{j=1}^{m} S_j = \sum_{j=1}^{m} W_j$$

where S_j denotes the $p \times p$ scatter matrix for the jth cluster and $\text{tr } S_j = W_j$. Equivalently, we have

$$W = \sum_{\ell=1}^{m} (\frac{1}{2n_\ell} \sum_{i=1}^{n_\ell} \sum_{j=1}^{n_\ell} d^2(X_i, X_j)) = \sum_{\ell=1}^{m} (\frac{1}{2n_\ell} \sum_{i=1}^{n_\ell} \sum_{j=1}^{n_\ell} d_{ij}^2) \qquad (3.1)$$

where $d^2(X_i, X_j) = (X_i - X_j)^T (X_i - X_j)$.

The purpose of a dynamic programming scheme is to systematically search for groupings which yield minimum values of the quantity W, eliminating those groupings which do not yield minimum values of W and also those that are redundant.

We now discuss the problem of partitioning $n = 6$ objects into $m = 3$ subsets by complete enumeration. This will serve to motivate a dynamic programming scheme for the cluster problem given by Jensen [183].

The total number of ways of partitioning 6 objects into 3 subsets is, by equation (2.11),

$$S(6,3) = \frac{1}{3!} \sum_{k=0}^{3} (-1)^k \binom{3}{k} (3-k)^6$$

$$= 90.$$

The 90 clustering alternatives can be classified according to their distribution forms [183]. The three distribution forms in this case are denoted by

$$
\begin{array}{llll}
\text{(i)} & \{4\} & \{1\} & \{1\}, \\
\text{(ii)} & \{3\} & \{2\} & \{1\}, \\
\text{(iii)} & \{2\} & \{2\} & \{2\},
\end{array}
$$

here each of the components in a distribution form $\{i\}$ denotes the number, i, of objects in the corresponding cluster. The components of a distribution form will always be written in decending order. In our example there are 90 clustering alternatives but only 3 distribution forms. In general the number of distribution forms is substantially smaller than the number of clustering alternatives.

There are $\dfrac{\binom{6}{4}\binom{2}{1}}{2} = 15$ clustering alternatives corresponding to the distribution form $\{4\}, \{1\}, \{1\}$; $\binom{6}{3}\binom{3}{2} = 60$ clustering alternatives corresponding to $\{3\}, \{2\}, \{1\}$; and $\binom{6}{2}\binom{4}{2}\binom{2}{2}/3! = 15$ clustering alternatives corresponding to $\{2\}, \{2\}, \{2\}$. The clustering alternatives corresponding to each distribution form are now listed.

Distribution Form $\{4\}, \{1\}, \{1\}$:

```
(1, 2, 3, 4), (5), (6)
(1, 2, 3, 5), (4), (6)
(1, 2, 5, 4), (3), (6)
(1, 3, 5, 4), (2), (6)
(5, 2, 3, 4), (1), (6)
(1, 2, 3, 6), (5), (4)
(1, 2, 6, 4), (5), (3)
(1, 6, 3, 4), (5), (2)
(6, 2, 3, 4), (5), (1)
(1, 2, 5, 6), (3), (4)
(1, 5, 6, 4), (2), (3)
(5, 6, 3, 4), (1), (2)
```

(5, 2, 3, 6), (1), (4)

(1, 5, 3, 6), (2), (4)

(5, 2, 6, 4), (1), (3)

Distribution Form {3}, {2}, {1}:

(1, 2, 3), (4, 5), (6)

(1, 2, 3), (4, 6), (5)

(1, 2, 3), (5, 6), (4)

(1, 2, 4), (3, 5), (6)

(1, 2, 4), (3, 6), (5)

(1, 2, 4), (5, 6), (3)

(1, 2, 5), (4, 3), (6)

(1, 2, 5), (4, 6), (3)

(1, 2, 5), (6, 3), (4)

(1, 2, 6), (4, 5), (3)

(1, 2, 6), (3, 5), (4)

(1, 2, 6), (3, 4), (5)

(1, 4, 3), (2, 5), (6)

(1, 4, 3), (2, 6), (5)

(1, 4, 3), (5, 6), (2)

(1, 5, 3), (4, 2), (6)

(1, 5, 3), (4, 6), (2)

(1, 5, 3), (2, 6), (4)

(1, 6, 3), (4, 5), (2)

(1, 6, 3), (4, 2), (5)

(1, 6, 3), (2, 5), (4)

(4, 2, 3), (1, 5), (6)

(4, 2, 3), (1, 6), (5)

(4, 2, 3), (5, 6), (1)

(5, 2, 3), (4, 1), (6)

(1, 4, 5), (2, 3), (6)

(1, 4, 5), (2, 6), (3)

(1, 4, 5), (3, 6), (2)

(1, 4, 6), (2, 3), (5)

(1, 4, 6), (2, 5), (3)

(1, 4, 6), (3, 5), (2)

(1, 5, 6), (2, 3), (4)

(1, 5, 6), (2, 4), (3)

(1, 5, 6), (3, 4), (2)

(2, 4, 5), (1, 3), (6)

(2, 4, 5), (1, 6), (3)

(2, 4, 5), (3, 6), (1)

(2, 4, 6), (1, 3), (5)

(2, 4, 6), (1, 5), (3)

(2, 4, 6), (3, 5), (1)

(2, 5, 6), (1, 3), (4)

(2, 5, 6), (1, 4), (3)

(2, 5, 6), (3, 4), (1)

(3, 4, 5), (1, 2), (6)

(3, 4, 5), (1, 6), (2)

(3, 4, 5), (2, 6), (1)

(3, 4, 6), (1, 2), (5)

(3, 4, 6), (1, 5), (2)

(3, 4, 6), (2, 5), (1)

(3, 5, 6), (1, 2), (4)

(5, 2, 3), (4, 6), (1)

(5, 2, 3), (1, 6), (4)

(6, 2, 3), (4, 5), (1)

(6, 2, 3), (4, 1), (5)

(6, 2, 3), (1, 5), (4)

(3, 5, 6), (1, 4), (2)

(3, 5, 6), (2, 4), (1)

(4, 5, 6), (1, 2), (3)

(4, 5, 6), (1, 3), (2)

(4, 5, 6), (2, 3), (1)

Distribution Form {2}, {2}, {2}:

(1, 2), (3, 4), (5, 6)

(1, 2), (3, 5), (4, 6)

(1, 2), (3, 6), (4, 5)

(1, 3), (2, 4), (5, 6)

(1, 3), (2, 5), (4, 6)

(1, 3), (2, 6), (4, 5)

(1, 4), (2, 3), (5, 6)

(1, 4), (2, 5), (3, 6)

(1, 4), (2, 6), (3, 5)

(1, 5), (3, 4), (2, 6)

(1, 5), (3, 2), (4, 6)

(1, 5), (3, 6), (2, 4)

(1, 6), (3, 4), (5, 2)

(1, 6), (3, 5), (4, 2)

(1, 6), (3, 2), (4, 5)

Under complete enumeration the objective function (WGSS) would need to be evaluated for each of the 90 clustering alternatives given above and that clustering alternative chosen for which W is a minimum. One notes from the list of clustering alternatives that under complete enumeration the WGSS would be calculated more than once for some of the clusters, for example, the cluster (1, 2, 3).

A dynamic programming scheme applied to the cluster problem is a scheme which works for the optimum grouping in stages such that at each stage the objective function is computed in such a way that redundant calculations inherent in the complete enumeration procedure are eliminated. In this way, the optimal solution will be attained in stages. The dynamic programming approach will require large amounts of rapid access storage.

The above example can be put into the framework of a dynamic programming solution as follows. The clustering alternatives are first classified according to their distribution forms. Recall that the distribution form components are listed in descending order. At the first stage the objective function for each cluster corresponding to the first distribution form component is evaluated and saved. At the second stage the objective function for the clusters corresponding to the first two components of the distribution forms is evaluated using all information from the first stage, that is, the within sum of squares is not recomputed for any cluster but "carried over" from the first stage.

For a discussion of the dynamic programming approach consider Table 3.1. The second column gives the clusters corresponding to the first componenet of the distribution forms, that is, the clusters available for the first stage. The number of clusters for the first stage is $\binom{6}{4}$ + $\binom{6}{3}$ + $\binom{6}{2}$ = 50. The function W will be computed for each of the fifty clusters in stage 1. At the second stage we will have 2 clusters corresponding to the first two components of the distribution forms, that is, we will have clusters of size {4} and {1}, {3} and {2}, or {2} and {2}. Thus the total number of objects at stage 2 will be 5 or 4. The number of ways of obtaining 5 objects is given by $\binom{6}{4}\binom{2}{1}$ + $\binom{6}{3}\binom{3}{2}$ = 90. The number of ways of obtaining 4 bjects is $\binom{6}{2}\binom{4}{2} + \binom{6}{3}\binom{3}{1} = 150$. The total number of ways of ob-

aining objects in stage 2 is therefore 240. However, there are

$\binom{6}{2}\binom{4}{2} = 45$ ways to form clusters giving rise to distribution form

components {2} {2} at stage 2, that is, there are 45 redundancies.

Also, for components {3}{1} at the second stage it will be necessary

to add 2 entities at the third stage giving rise to distribution form

{3} {1} {2} which is ultimately equivalent to form {3} {2} {1}. Thus,

the total number of ways of obtaining entities for the second stage is

$$\binom{6}{4}\binom{2}{1} + \binom{6}{3}\binom{3}{2} + \frac{1}{2}\binom{6}{2}\binom{4}{2} = 30 + 60 + 45 = 135,$$

a reduction of 105.

The number of distinct sets containing either 4 or 5 objects for

stage 2 is $\binom{6}{5} + \binom{6}{4} = 21$. These are listed under stage 2 in Table

3.1 and are called states. Thus there are 21 states in stage 2.

There were 50 states in stage 1. Five of the 135 <u>feasible ways</u> of ob-

taining states in stage 2 are indicated in Table 3.1.

The final stage of the process is stage 3. The final stage will

result in 3 clusters. There is only one state in the final stage, the

one involving all six objects. The number of ways of arriving at the

six objects in the final stage is

$$\binom{6}{5}\binom{1}{1} + \binom{6}{4}\binom{2}{2} = 6 + 15 = 21,$$

that is, there are 21 feasible arcs from stage 2 to stage 3.

For example with n = 6 and m = 3 there are a total of 135 + 21 =

156 feasible arcs. If one includes the number of initial states then

there are 156 + 50 = 206 feasible arcs.

50

TABLE 3.1

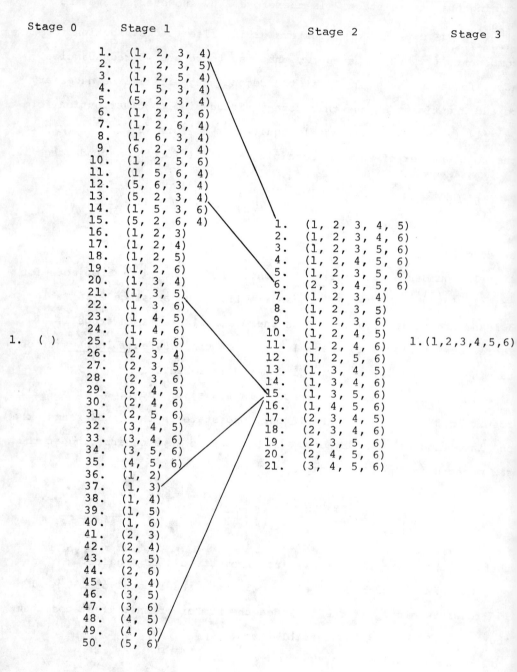

ach feasible arc results in what is called a <u>transitional calculation</u>

efined by

$$T(g_k) = \frac{1}{n_k} \sum_{i<j \varepsilon g_k} d_{ij}^2 \qquad (3.2)$$

where g_k denotes a group of n_k objects and d_{ij} the distance between X_i and X_j.

The total enumeration procedure involves 90 clustering alterna-

tives and 3 transitional calculations for each alternative resulting

in a total of 270 transitional calculations. The dynamic programming

approach involves 206 or 64 fewer transitional calculations.

Under dynamic programming suppose there exists a state, at some

stage k, containing objects $X_1, \ldots, X_q$, $q \leq n$. The dynamic programming

procedure stores in memory the optimal way to partition the q objects

in k nonempty and mutually exclusive subsets. In later stages in

which the q objects are partitioned into k subsets it is not necessary

to recompute all feasible ways of performing the partitioning.

As an illustration consider our example with n = 6, m = 3.

TABLE 3.2

Alternative	Transitional Calculations
1	T(1, 2) + T(3, 4) + T(5, 6)
2	T(1, 3) + T(2, 4) + T(5, 6)
3	T(1, 4) + T(2, 3) + T(5, 6)

Recall that when n = 6 and m = 3 there are S(6, 3) = 90 unique cluster-

ing alternatives available. Three of these are listed in Table 3.2.

Under complete enumeration 9 transitional computations would be re-

quired for these 3 alternatives. Under dynamic programming it would take 6 transitional computations for the optimal partition of (1, 2, 3, 4) into two groups of size 2. The optimal partition, say $T(1, 3) + T(2, 4) = W_2(1, 2, 3, 4)$ is recorded in memory so that only one additional computation is required to determine $W_2(1, 2, 3, 4) + T(5, 6)$. For these 3 alternatives dynamic programming has eliminated $9 - 7 = 2$ redundant calculations. Actually, as n and m are increased the number of redundant arcs that are eliminated is substantial, however, relative to the total number of transitional calculations the difference may not be so great.

3.2. Jensen's Dynamic Programming Model

There is no standard mathematical formulation for the dynamic programming problem. This is in contrast to the linear programming problem for which there does exist a precise standard formulation. The equations and formulas pertinent to a dynamic programming problem depend on the particular situation at hand. The problem is usually reduced to a recursive relationship or equation which reflects the multiple interrelated decisions inherent in the dynamic programming procedure and which result in the final "optimal" result. Rao [291] gives a dynamic programming formulation for the cluster problem when $p = 1$. Jensen [183] gives a more general formulation which we now consider.

Jensen's dynamic programming formulation [183] is given in terms of the recursive equation

$$W_k(z) = \begin{cases} 0 & \text{if } k = 0 \\ \\ \min_y [T(z-y) + W_{k-1}(y)], & \text{if } k = 1,2,\ldots,m_0. \end{cases} \tag{3.3}$$

where

    $m \equiv$ number of disjoint and non-empty subsets into which the n objects are to be partitioned,

    $k \equiv$ index or stage variable,

    $m_o \equiv m$ if $n \geq m$, and $n - m$ if $n < m$,

    $z \equiv$ state variable representing a given set of objects at stage k,

    $y \equiv$ state variable representing a given set of objects at stage k - 1,

    $z - y \equiv$ subset of all objects contained in z but not in y,

    $T(z-y) \equiv$ is the "transition cost" of the objects in the cluster of objects in (z-y).

The variables y and z represent 2 states (sets of objects) in stages k - 1 and k, respectively. The difference z - y represents those objects contained in the stage k state but not in the stage k - 1 state. $T(z-y)$ then represents the "transition cost" or WGSS for those objects which are combined with the stage k - 1 state objects and $W_k(z) =$ min $[T(z-y) + W_{k-1}(y)]$ gives the minimum value for WGSS in partition-
  y
ing the objects represented by z into k disjoint and nonempty subsets. It will be seen that the use of formula (3.3) calls for a substantial amount of bookkeeping. Recall from section 3.1, eq.(3.2), that if g_i denotes a cluster of n_i objects then the transition cost $T(g_i)$ is given by

$$T(g_i) = \frac{1}{n_i} \sum_{k<j \epsilon g_i} d_{kj}^2$$

which is actually the WGSS for cluster g_i.

Note that the number of stages is $m_0 = m$ if $n \geq m$, and $n - m$ if $n < 2m$. The reason for this is that if $n < 2m$ there must always be at least $n - m + 1$ single-object clusters. The transition cost T for a single-object cluster is 0 so that single-object clusters add nothing to W. Consequently, the process may be terminated at stage m_0 and all remaining clusters are assumed to be single-object clusters. Also, in computing $W_k(z)$ it should be emphasized that the objects corresponding to any state in stage k consist of objects contained in some set corresponding to some state y of stage $k - 1$ and objects contained in another set represented by $z - y$.

As an example to illustrate the notions involved in the recursive equation (3.3) consider state 37 of stage 1 and state 15 of stage 2 when $n = 6$ and $m = 3$ (Table 3.1). In this case y represents the objects (1, 3) in stage 37 of stage 1, z represents the objects (1, 3, 5, 6) in state 15 of stage 2, and $z - y$ represents the objects (5, 6). The "transition cost" from stage 37 to state 15 is then

$$T(z - y) = T(5, 6) = d_{56}^2$$

The transition cost from state 37 in stage 1 to state 1 in stage 2 would be

$$T(z - y) = T(2, 4, 5) = \frac{d_{24}^2 + d_{25}^2 + d_{45}^2}{3}$$

At the first stage the dynamic programming algorithm considers the evaluation of $W_1(z)$ for a given set of clusters. In this case

$$W_1(z) = \min_{y} \; [T(z-y) + W_0(y)] = T(z),$$

where z represents a given set of objects. The quantity $W_1(z)$ is computed for each of the possible clusters at the first stage. The maximum number of objects available for a cluster in the first stage, denoted by max (1), is given by

$$\text{max } (1) \equiv n - m + 1,$$

that is, the largest cluster has $n - m + 1$ objects in which case the remaining clusters would be single-object clusters. The minimum number of objects, denoted by min (1), in a cluster in stage 1 is

$$\text{min } (1) = n/m$$

if n is an even multiple of m, and

$$\text{min } (1) = \begin{cases} [n/m] + 1, & \text{for } 1 \leq n - m[n/m], \\ \\ n - (m-1) \ [n/m], & \text{for } n - m[n/m] < 1 \leq m, \end{cases}$$

when n is not an even multiple of m, where $[n/m]$ denotes the largest integer $\leq n/m$. The total number of clusters available for the first stage, denoted by NS(1) is given by

$$\text{NS} (1) = \sum_{j=\text{min}(1)}^{\text{max}(1)} \binom{n}{j}. \tag{3.4}$$

The first stage of the algorithm consists of computing the quantity $T(z)$ for each of the NS(1) possible clusters.

In general the maximum number of objects in any one state in stage k is equal to the maximum sum of distribution form components

from stages 1 through k inclusive. The minimum number of states is
the minimum sum of the distribution form components. For max(k) and
and min(k) we have

$$\max(k) = n - m + k \qquad (3.5)$$

and

$$\min(k) = k[n/m] \qquad (3.6)$$

if n is an even multiple of m. If n is not an even multiple of m we
have

$$\min(k) = \begin{cases} ([n/m] + 1)k, & \text{for } 1 \le k \le n - m[n/m] \\[2ex] n - (m - k)[n/m], & \text{for } n - m[n/m] < k \le m. \end{cases} \qquad (3.7)$$

The number of states available for stage k is given by

$$NS(k) = \begin{cases} 1 & \text{for } k = 0 \\[2ex] \sum_{j=\min(k)}^{\max(k)} \binom{n}{j} & \text{for } k = 1,2,\ldots,m_0. \end{cases} \qquad (3.8)$$

The total number of states available in the dynamic programming
formulation is thus given

$$\sum_{k=0}^{m_0} NS(k). \qquad (3.9)$$

A very important quantity in the formulation is the total number of values for $W_k(z)$ in going from stage $k - 1$ to stage k, that is, the number of ways of forming a state in stage k. States in successive stages are connected by arcs. Two states, in stages $k - 1$ and k, are connected by a feasible arc if the objects in the state in stage k are also contained in the state in stage $k-1$. That is a feasible arc cannot exist between a state in the stage $k - 1$ and a state in stage k if an object contained in the stage $k - 1$ state is not contained in the stage k state for $2 \leq k \leq m_0$.

In the dynamic programming algorithm the total number of feasible arcs is given by

$$TFA \equiv NS(1) + \sum_{k=1}^{m_0-1} TA(k) \tag{3.10}$$

where $TA(k)$ represents the total number of feasible arcs between stage k and stage $k+1$ for $k = 1, 2, \ldots, m_0$. The value of $TA(k)$ is given by

$$TA(k) = \sum_{j=min(k)}^{max(k)} \sum_{i=1}^{max(k+1)-min(k)} FA(j,i), \tag{3.11}$$

where

$$FA(j,i) = \begin{cases} \binom{n}{j} \binom{n-j}{i} & \text{if } min\ (k+1) \leq i + j \leq max(k+1) \\ \\ 0 & \text{otherwise.} \end{cases} \tag{3.12}$$

In (3.11) and (3.12), i denotes the number of objects among a class of (feasible) <u>states</u> at stage k. There are $\binom{n}{i}$ such states containing i objects, since $\binom{n}{i}$ is the number of subsets of size i. The quantity j denotes the number of objects to be combined with the i objects to form a new state at stage k+1. Obviously we must have $\min(k+1) \leq i + j \leq \max(k+1)$ for a state of size $i + j$ to exist at stage k+1. If $i + j$ satisfies the required condition then there are $\binom{n-i}{j}$ sets of size n-i that may be added to the i objects j at a time.

Jensen gives a way of computing the efficiency of dynamic programming relative to complete enumeration. Efficiency is defined as the ratio of the total number of transitional calculations under dynamic programming to the corresponding number of calculations under complete enumeration. Alternatively, the numerator can be taken to be the total number of feasible arcs. In either case the dynamic programming procedure is quite efficient. However, the dynamic programming procedure requires more computer memory and consequently slow-access storage could make it less useful than complete enumeration. In any event for large n and m one might be better off using some other technique such as ISODATA [18] or hierarchial procedures.

In order to illustrate Jensen's formulation consider the example with n = 6 and m = 3. In this case n = 2m so we need to consider $n - m = 6 - 3 = 3$ stages, i.e. $m_0 = 3$. Furthermore,

$$\max(1) = n - m + 1 = 4$$
$$\min(1) = ([6/3])1 = 2$$
$$\max(2) = n - m + 2 = 5$$
$$\min(2) = ([6/3])2 = 4$$
$$\max(3) = n - m + 3 = 6$$
$$\min(3) = ([6/3])3 = 6$$

as can be seen from Table 3.1.

The total number of states in stages 0, 1, 2, and 3 are given by

$$NS(0) = 1$$

$$NS(1) = \binom{6}{4} + \binom{6}{3} + \binom{6}{2} = 50$$

$$NS(2) = \binom{6}{4} + \binom{6}{5} = 21$$

$$NS(3) = \binom{6}{6} = 1$$

These states are listed in Table 3.1. The total number of states is thus 73. This figure agrees with Table 3.1 if $NS(0) \equiv 1$.

From equation (3.12) we have, for k=1,

$$FA(3,1) = \binom{6}{3}\binom{3}{1} = 60$$

$$FA(3,2) = \binom{6}{3}\binom{3}{2} = 60$$

$$FA(4,1) = \binom{6}{4}\binom{2}{1} = 30$$

$$FA(2,2) = \binom{6}{2}\binom{4}{2} = 90$$

$$FA(3,3) = FA(4,2) = 0$$

and the total number of feasible arcs between stage 1 and stage 2 is

$$TA(1) = 240.$$

Similarly for k=2 we have

$$TA(2) = \binom{6}{4}\binom{2}{2} + \binom{6}{5}\binom{1}{1} = 21.$$

Thus the total number of feasible arcs in our example is, by (3.10),

$$TFA = NS(1) + \sum_{k=1}^{2} TA(k) = 50 + 240 + 21 = 311.$$

From section 3.1 it was seen that half of the states in stage 2 corresponding to distribution form components {2}, {2} are redundant and that the 60 arcs corresponding to components {3} {1} ultimately lead to the form {3} {1} {2} which is equivalent to {3} {2}{1}. Thus in the reduced formulation the number of feasible arcs, denoted by NA,

$$NA = 50 + 135 + 21 = 206.$$

The number of feasible arcs, between stages k and k+1, after redundance elimination is given by

$$NA(k) = \sum_{i=\min(k)}^{\max(k)} \sum_{j=1}^{\max(k+1)-\min(k)} A(i,j)$$

where

$$A(i,j) = \begin{cases} \binom{n}{i}\binom{n-i}{j} & \text{if } i \neq j \\ \frac{1}{2}\binom{n}{i}\binom{n-i}{j} & \text{if } i = j \\ 0 & \text{otherwise,} \end{cases} \quad \text{and} \quad \begin{cases} \min(k+1) \leq i+j \leq \max(k+1) \\ (m-k)j + i \geq n \end{cases}$$

The total number of arcs in the reduced formulation is given by

$$NA = NS(1) + \sum_{k=1}^{m_0-1} NA(k). \qquad (3.13)$$

It can be verified that (3.13) yields 206 when n = 6 and m = 3. The maximum number of feasible arcs that must be considered in the dynamic programming formulation in this case is then 206.

To illustrate how the dynamic programming algorithm operates let p=2 and let the six objects be (1,1), (3,4), (5,5), (4,4), (1,2), and (5,6) or

$$X = \begin{pmatrix} 1 & 3 & 5 & 4 & 1 & 5 \\ 1 & 4 & 5 & 4 & 2 & 6 \end{pmatrix}$$

The squared distances are then

$$d_{12}^2 = 13, \quad d_{13}^2 = 32, \quad d_{14}^2 = 18, \quad d_{15}^2 = 1, \quad d_{16}^2 = 41,$$

$$d_{23}^2 = 5, \quad d_{24}^2 = 1, \quad d_{25}^2 = 8, \quad d_{26}^2 = 8, \quad d_{34}^2 = 2,$$

$$d_{35}^2 = 25, \quad d_{36}^2 = 1, \quad d_{45}^2 = 13, \quad d_{46}^2 = 5, \quad d_{56}^2 = 32.$$

According to the dynamic programming algorithm we would have:

Stage 0: $\qquad\qquad\qquad W_0(0) = 0.$

Stage 1: Compute $W_1(z) = T(z-y) + W_0(y) = T(z) + W_0(0) = T(z)$

for each set of objects in stage 1. For example

$$W_1(1, 2, 3, 4) = T(1, 2, 3, 4) + W_0(0)$$

$$= \frac{(d_{12}^2 + d_{13}^2 + d_{14}^2 + d_{23}^2 + d_{24}^2 + d_{34}^2)}{4}$$

$$= 17.75$$

There are 50 such values for stage 1.

Stage 2: Compute $W_2(z) = \min_{y} \{T(z-y) + W_1(y)\}$ for each set of objects in stage 2. For example

$$W_2(1, 2, 3, 4, 5) = \min \{T(5) + W_1(1, 2, 3, 4), \ T(4) + W_1(1, 2, 3, 5)$$

$$T(3) + W_1(1,2,4,5), \quad T(2) + W_1(1,3,4,5),$$

$$T(1) + W_1(2,3,4,5), \quad T(1,2) + W_1(3,4,5),$$

$$T(1,3) + W_1(2,4,5), \quad T(1,4) + W_1(2,3,5),$$

$$T(1,5) + W_1(2,3,4), \quad T(2,3) + W_1(1,4,5),$$

$$T(2,4) + W_1(1,3,5), \quad T(2,5) + W_1(1,3,4),$$

$$T(3,4) + W_1(1,2,5), \quad T(3,5) + W_1(1,2,4),$$

$$T(4,5) + W_1(1,2,3)\}.$$

Stage 3: Compute $W_3(z) = \min_y\{T(z-y) + W_2(y)\}$ for each set of objects in stage 3. In this stage z represents the one set of objects (1,2,3,4,5,6). There are 21 feasible arcs between states in stage 2 and states in stage 3. Thus, we would choose the minimum of 21 values. As an example one of these 21 values is

$$T(2,4) + W_2(1,3,5,6)$$

corresponding to state number 15 (see Table 3.1) in which case y corresponds to the set (1,3,5,6), z corresponds to the set (1,2,3,4,5, 6) and z-y corresponds to the set (2,4).

The results of the dynamic programming procedure are the clusters (1,1) and (1,2); (3,4) and (4,4); and (5,5) and (5,6) with distribution form {2}, {2}, {2}. The minimum value for W is

$$W_3(1,2,3,4,5,6) = 1.5$$

The results are displayed in Figure 3.1.

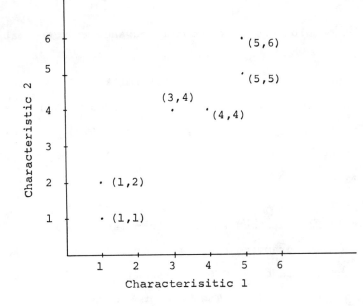

Figure 3.1.   Graph of n=6 objects

3.3   Integer Programming Applications to Cluster Analysis

In this section the cluster problem will be considered in terms
of integer programming formulations due to Vinod [380] and Rao [291].
Balinsky [12] surveys some advances made in Integer Programming.
Other pertinent references on Integer Programming are [11], [26],
[134], and [228].

A formulation due to Vinod [380] considers a measurable characteristic (p=1) with respect to which the clustering is to be performed.
Suppose we are given n objects with values $x_1, \ldots, x_n$. Let n_j denote
the number of objects in the jth group (j = 1,...,m) with $n = \sum_{j=1}^{m} n_j$.
The size of the largest cluster will be m_0. If m_0 is unspecified then
m_0 = n. The cost involved in placing the ith object in the jth group
will be denoted by c_{ij}, with $c_{ii} = 0$ and $c_{ij} = c_{ji}$. Now let $a_{ij} = 1$

or 0 according as the ith object is or is not contained in the jth group. We have $n_j = \sum_{i=1}^{n} a_{ij}$, that is n_j is the sum of the elements in the jth column of $A = \{a_{ij}\}$.

If the number of clusters m is known in advance then the matrices $C = \{c_{ij}\}$ and $A = \{a_{ij}\}$ would be of order $m \times n$. However, if m is not known in advance then the order of C and A is not known. To circumvent this arrangement we assume there are a total n groups with n - m empty groups. For the purpose of identifying the groups the idea of group leader is created and the jth object is taken to be the leader of the jth group. Let $y_j = 1$ if the jth object is a leader and $y_j = 0$ otherwise.

The cost matrix C is assumed known in formulating the cluster problem as an integer programming problem. For example c_{ij} could be considered to be the increase in WGSS in placing object i into the group that has j as group leader. The Integer Programming problem is to minimize the total cost of any scheme of clustering subject to certain constraints, that is,

$$\text{minimize} \quad \sum_{i=1}^{n} \sum_{j=1}^{n} a_{ij}\, c_{ij}$$

subject to

(i) $\quad \sum_{j=1}^{n} a_{ij} = 1, \quad i = 1,2,\ldots,n,$ \hfill (3.14)

(ii) $\quad \sum_{j=1}^{n} y_j = m$ \hfill (3.15)

(iii) $y_j \geq a_{1j},\; y_j \geq a_{2j},\ldots,y_j \geq a_{nj},\; j = 1,2,\ldots,n.$ \hfill (3.16)

Constraint (i) states that any object cannot be put into more than one group. Constraint (ii) states that there are exactly m groups. The last constraint (iii) says that the jth object must be a leader before any objects can be placed in the group corresponding to it. Constraint (iii) can be stated as

(iii)'

$$ny_j \geq \sum_{i=1}^{n} a_{ij}, \quad j = 1,2,\ldots,n.$$

The above formulation can be applied to the case where each object is a p-dimensional vector, that is,

$$x_i^T = (x_{1i}, x_{2i},\ldots,x_{pi}).$$

The cost in this case could be defined to be any one of the distance metrics defined in chapter 1. Once this has been done the dimensionality of the cost matrix C remains the same. The problem then is to minimize C with respect to the new definition of C.

In a second formulation, due to Vinod [380], the cost c_{ij} is defined in terms of WGSS. The values $x_1,x_2,\ldots,x_n$ of the n objects are ordered in increasing magnitude, and are denoted by $z_1,z_2,\ldots,z_n$, i.e.,

$$z_1 < z_2 < \ldots < z_n$$

The problem is then one of partitioning $z_1,z_2,\ldots,z_n$. Let g_j denote the group for which the smallest quantity is z_j. The quantity z_j is the group leader for group g_j. The matrix A is defined in the same fashion, that is, $a_{ij} = 1$ or 0 as z_i belongs or does not belong to g_j. Because of the order property of the z's, z_i cannot belong to g_{i+1}, $\ldots,g_n$ which means the elements of A above the diagonal will be zero, or equivalently

$$\sum_{i=1}^{n} \sum_{j=i+1}^{n} a_{ij} = 0$$

The problem can be stated as one of minimizing

$$\sum_{i=1}^{n} \sum_{j=1}^{n} a_{ij} (z_i - \bar{z}_j)^2, \tag{3.17}$$

where $\bar{z}_j$ is the mean of the jth group,

$$\bar{z}_j = \sum_{i=1}^{n} a_{ij} z_i / n_j$$

and

$$\sum_{i=1}^{n} a_{ij} = n_j.$$

A condition necessary for (3.17) to be a minimum is the so-called string property. The string property states that there can be no group containing z_i and z_j ($i < j$) without containing all the values between z_i and z_j, which implies that the matrix $A = \{a_{ij}\}$ cannot have an interrupted string of ones below the diagonal which can be stated as

$$a_{jj} \geq a_{j+1}, \ j \geq \cdots \geq a_{nj}, \ j = 1,2,\ldots,n.$$

The longest string can have m_0 ones, since m_0 is the most objects that can belong to a group.

Lemma 3.1. The string property $a_{jj} \geq a_{j+1}, \ j \geq \cdots \geq a_{nj}$, is necessary for the minimization of $\sum_{i=1}^{n} \sum_{j=1}^{n} a_{ij}(z_i - \bar{z}_j)^2$.

Proof: The increase in WGSS in combining the objects corresponding to $z_j, z_{j+1}, \ldots, z_{j+k-1}$ with that corresponding to z_{j+k} is, according to the trace of (1.10),

$$\frac{k}{k+1} \left(z_{j+k} - \frac{1}{k} \sum_{i=0}^{k-1} z_{j+1}\right)^2.$$

The similar result for $z_j, z_{j+1}, \ldots, z_{j+k-1}$ and z_{j+k+1} is

$$\frac{k}{k+1} \left(z_{j+k+1} - \frac{1}{k} \sum_{i=0}^{k-1} z_{j+i}\right)^2$$

However, $z_{j+k+1} > z_{j+k}$ which proves the desired result.

Now, the minimization of $\sum_{i=1}^{n} \sum_{j=1}^{n} a_{ij} (z_i - \bar{z}_j)^2$ can be stated as

the minimization of $\sum_{i=1}^{n} \sum_{j=1}^{n} a_{ij} c_{ij}$ with the appropriate choices of

$= \{a_{ij}\}$ and $C = \{c_{ij}\}$, C being defined in such a way that

$$\sum_{i=1}^{n} \sum_{j=1}^{n} a_{ij} c_{ij} = \sum_{i=1}^{n} \sum_{j=1}^{n} a_{ij} (z_i - \bar{z}_j)^2$$

The quantities c_{ij} need to be defined only for $i > j$, that is only for elements below the main diagonal. The value c_{ij} will be defined to be the increase in WGSS when z_i is included in the group having objects corresponding to $z_j, \ldots, z_{i-1}$. We have, according to (1.10),

$$c_{ij} = \frac{i-j}{i-j+1} \left(z_i - \frac{1}{i-j} \sum_{k=j}^{i-1} z_k\right)^2. \tag{3.18}$$

For this particular choice of c_{ij} it can be shown that

$$\sum_{i=1}^{n} \sum_{j=1}^{n} a_{ij} \, c_{ij} = \sum_{i=1}^{n} \sum_{j=1}^{n} a_{ij} \, (z_i - \bar{z}_j)^2$$

which reduces the original quadratic programming problem to a linear integer programming problem. The problem can finally be stated as:

$$\text{minimize} \quad \sum_{i=1}^{n} \sum_{j=1}^{n} a_{ij} \, c_{ij}$$

subject to

(i) $\quad a_{jj} > a_{j+1,j} > \cdots > a_{nj}, \; j = 1,2,\ldots,n,$

(ii) $\quad c_{ij} = \dfrac{i-j}{i-j+1} \, (z_i - \dfrac{1}{i-j} \sum_{k=j}^{i-1} z_k)^2.$

The extension to the multivariate case of Vinod's second formulation is not as straight forward as for his first formulation. Let $X_1, X_2,\ldots,X_n$ denote n points in p-dimensional space. Since the string property associated with the univariate case cannot be used with the multivariate case, Vinod [380] defines a (generalized) string property associated with multivariate case. Furthermore, he states that the string property is necessary for the minimization of WGSS, a conjecture to which Rao [291] gives a counter example. Rao proceeds to give an alternative definition for generalized string property which is also not necessary for the minimization of WGSS.

The two definitions of generalized string property will be given after which the two versions of the problem (Rao's and Vinod's) are briefly stated and discussed. Let

$$D = \{d_{ij}\}$$

be the n × n matrix of pairwise Euclidean distances. The quantities

n the jth column of D can be reordered in increasing order of magni-
ude, that is,

$$d_{jj} < d_{i_1, j} < \cdots < d_{i_{n-1}, j}$$

hich is analogous to ordering the z's in the univariate case. In
erms of the matrix A the generalized string property states that a
tring of ones should now follow the pattern given by $i_1, i_2, \ldots, i_{n-1}$.
he maximum number of values for any column is m_0.

The two definitions of generalized string property are

(1) (Vinod): The string of ones in the jth column of
A should follow the pattern given by $i_1, i_2, \ldots, i_{n-1}$
according to

$$d_{jj} = 0 \leq d_{i_1, j} \leq d_{i_2, j} \leq \cdots \leq d_{i_{n-1}, j}.$$

(2) (Rao): Each group consists of at least one object
(group leader) such that the distance between the
leader and any object that does not belong to the
same group is not less than the distance between
the leader and any object within the group.

Vinod defines $c_{i_k, j}$ ($i_k \neq j$) to be the increase in WGSS when X_{i_k}
is placed in the group containing objects j, $i_1, \ldots, i_{k-1}$. The quantity $c_{i_k, j}$ corresponding to (3.18) in the univariate case is given by

$$c_{i_k, j} = \frac{k}{k+1} (d^2_{i_k, j} - \frac{1}{k} \sum_{m=1}^{k} d_{i_m, j})^2 \qquad (3.19)$$

Using the cost matrix corresponding to (3.19) and taking the A matrix to satisfy $a_{ij} \geq a_{i_1,j} \geq a_{i_2,j} \geq \cdots \geq a_{i_{n-1},j}$ the problem of minimizing WGSS is reduced to that of minimizing $\sum\limits_{i=1}^{n} \sum\limits_{j=1}^{n} a_{ij} c_{ij}$.

Rao [291] requires string property (2) to be satisfied in his linear integer programming formulation. The problem is to determine

$$\min \; c^T Y \qquad\qquad (3.20)$$

subject to

$$A \; Y = b^T, \text{ every } y_i = 0 \text{ or } 1 \text{ except the last,}$$

where

A is a $(n + 1) \times [n(n - 1) + 2]$ matrix,

Y is a $n(n - 1) + 2$ element column vector,

b is a $n + 1$ element column vector given by $(1,1,\ldots,1,m)^T$,

c^T is $n(n - 1) + 2$ element row vector.

Note that the quantity to be minimized in (3.20) is again a linear function of 0-1 variables. Each element of Y represents a particular potential group, that is, $y_i = 0$ or 1 according as the particular group (ith group) is or is not utulized in the final solution. The vector c^T is a vector of objective function values, that is, c_i represents the objective function value for the particular ith group. For example, if the objective function is WGSS, then c_i is of the form

$$c_i = \sum_{\ell=1}^{n_i} \sum_{j=1}^{p} (x_{j\ell} - \bar{x}_j)^2.$$

The vector Y contains m ones and $n(n-1) + 2 - m$ zeros. The quantity $^T Y$ then yields the objective function value for a particular clustering alternative.

The matrix A is analogous to the one in (3.17), however, now its order is $(n + 1) \times [n(n - 1) + 2]$. The matrix A satisfied the following conditions:

(i)   the last row of A has all ones, that is,

$a_{n+1,j} = 1$, $j = 1,2,\ldots,n(n-1) + 2$,

(ii) each row of A, except the last, corresponds to an object, and has only a single one and $n(n-1) + 1$ zeros,

(iii) each column of A, except the last, represents the coefficients for a particular group, that is

$$\sum_{i=1}^{n} a_{ij} = n_j, \quad j = 1,2,\ldots,n(n-1) + 1$$

(iv) the last column of A has all zeros except for a 1 in the last row.

The last element of b insures that the total number of clusters is m.

It will be assumed that $d_{ij} \neq d_{ik}$ if $j \neq k$. As a consequence of the string property it follows that there are $n - 1$ potential groups that have object j as leader. To see this consider

$$d_{jj} = 0 < d_{i_1,j} < d_{i_2,j} < \cdots < d_{i_{n-1},j}.$$

With object j as leader the following groupings are possible; (j,i_1), $(j,i_1,i_2)\ldots,(j,i_1,i_2,\ldots,i_{n-1})$. There are n-1 of these. Since there are n objects it follows that the total number of possible groups is n(n-1) + 1 including the one consisting of all objects. The number of elements in the vector Y corresponds to the total number of possible clusters and the restriction that the total number of possible clusters is m. The problem given by (3.20) and its associated constraints can be solved by a set partitioning technique given in [127].

Rao gives three other criteria for minimization that can be used which lead to problems in mathematical programming. These are:

(i)   minimization of the sum of the average within group
      squared distances;

(ii) minimization of the total within groups distance;

(iii) minimization of the maximum within group distances.

The solutions given in these three cases are not very useful, computationally, unless m and n are small. When the number of groups m = 2 the solutions become more feasible. This is a popular value for m in view of the Edwards and Covalli-Sforza [93] technique which divides the objects into two compact groups and repeats the procedure sequentially.

Chapter 4

SIMILARITY MATRIX REPRESENTATIONS

As we have seen in the previous chapters, cluster analysis can be discussed either in terms of the distance matrix D of all inter-point distances among the n objects or the similarity matrix S. In this chapter we will discuss some aspects regarding the representation of clustering results or of similarity or distance matrices. The presentation in sections 4.1 and 4.2 is rather informal leading up to a more precise formulation (due to Hartigan) in sections 4.3 and 4.4.

4.1. Dendograms

In an hierarchical clustering process one starts with n objects and groups the two nearest (or most similar) objects into a cluster thus reducing the number of clusters to n - 1. The process is repeated until all objects have been grouped into the cluster containing all n objects. The ideas presented in this chapter will be concerned with hierarchical (numerically stratified) methods.

The most commonly used method of representing a distance or similarity matrix is perhaps the method which utilizes the idea of "dendogram" or "tree diagram". The dendogram may be viewed as a diagrammatic representation of the results of an hierarchical clustering process which is carried out in terms of the distance or similarity matrix. Now a hierarchical clustering procedure can be considered to be a procedure which operates on the distance or similarity matrix. The dendogram is then a structure that depicts geometrically or graphically the hierarchical clusters that result when a given clustering procedure operates on a similarity or distance matrix.

There are several ways to draw a tree diagram corresponding to a given dendogram. In a tree diagram the objects are labeled vertically on the left and the clustering results are indicated to the right. The levels of similarity or distance at which new clusters are formed are indicated horizontally across the top of the diagram. Given n objects there are many possible trees that can be formed by the hierarchical procedure, however, given a particular distance or similarity matrix, there corresponds to it only one tree diagram.

Figure 4.1 shows an example of a tree diagram. We will consider tree diagrams of this type only so that we may think of tree diagrams and dendograms as being one and the same. Figure 4.1 corresponds to the case of six objects (n=6) and p characters (variates). The objects 1 and 3 are most similar (closest) and are thus clustered together at the similarity level of .9.

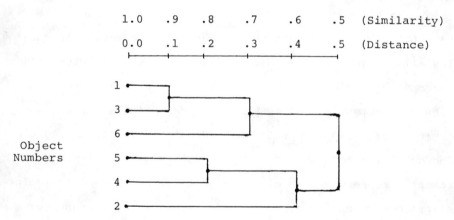

Figure 4.1

Objects 4 and 5 are then grouped at level .8. At this stage of the process there are 4 clusters, (1,3), (6), (5,4), and (2). At the third and fourth stages the process forms clusters (1,3,6) and (5,4,2) at levels of similarity of .7 and .6, respectively. Finally, all objects are combined to form the cluster of all objects at level .5. For some comments on representation of the results of clustering, see Sokal and Sneath [336]. Given n objects, the construction of the dendogram depends on the choice of similarity coefficient or distance between two single-object clusters and a clustering technique. More important, however, is the choice of similarity or distance between two clusters of objects. Several measures of distance have been discussed in chapter 1. The construction of a dendogram could be carried out by using any one of these measures.

Johnson [186] considers hierarchical clustering schemes and develops a correspondence between such schemes and a very special type of metric. Consider the sequence of clusterings C_0, C_1, C_2, ...,C_m and associate with each a number α_j, $j = 0,1,...,m$. In terms of the example illustrated in Figure 4.1 there is a clustering C_j associated with each branch point and α_j is the corresponding level at which that clustering was performed. The clustering C_0 is that containing n single-object clusters and $\alpha_0 = 0$. Thus we have $\alpha_0 \leq \alpha_1 \leq \cdots \leq \alpha_m$ and every cluster in C_j is the union of clusters in C_{j-1}. Such an arrangement is termed a hierarchical clustering scheme (HCS).

Johnson further demonstrates that every HCS yields a special kind of metric between the n objects, and conversely that the HCS can be recovered given such a metric. Thus one may study the HCS's by studying the corresponding metrics.

Given a HCS, C_0, C_1,...,C_m with values α_0, α_1,...,α_m the distance measure $d(X_p, X_q)$ is defined by

$$d(X_p, X_q) = \alpha_i, \qquad (4.1)$$

where i is the smallest integer in the set $\{0,1,2,...,m\}$ such that

$X_p \; \varepsilon \; C_i$ and $X_q \; \varepsilon \; C_i$. For example in terms of Figure 4.1 we have $\dot{d}(X_2, X_5) = .4$ and $d(1,6) = .3$. The distance matrix corresponding to the measure d is

$$D = \begin{pmatrix} 0 & .5 & .1 & .5 & .5 & .3 \\ .5 & 0 & .5 & .4 & .4 & .5 \\ .1 & .5 & 0 & .5 & .5 & .3 \\ .5 & .4 & .5 & 0 & .2 & .5 \\ .5 & .4 & .5 & .2 & 0 & .5 \\ .3 & .5 & .3 & .5 & .5 & 0 \end{pmatrix} \qquad (4.2)$$

It can be shown that the distance measure (4.1) is a bona fide metric (see definition 1.1). Of particular interest is the verification of the triangle inequality. To verify it let X, Y, and Z be any three objects with $d(X,Y) = \alpha_j$ and $d(Y,Z) = \alpha_k$. This implies X and Y are in some cluster contained in C_j, and Y and Z are in some cluster contained in C_k. However, the cluster contained in C_i, where i = max (j,k), contains the other cluster, since we are dealing with a HCS. Thus X, Y, and Z are in the same cluster contained in C_i. Consequently

$$d(X,Y) \leq \alpha_i = \max (\alpha_j, \alpha_k) \; ,$$

$$d(Y,Z) \leq \alpha_i = \max (\alpha_j, \alpha_k) \; ,$$

and

$$d(X,Z) \leq \max [d(X,Y), \; d(Y,Z)]. \qquad (4.3)$$

Inequality (4.3) is called the <u>ultrametric inequality</u>. Since max $[d(X,Y), \; d(Y,Z)] \leq d(X,Y) + d(Y,Z)$ it is stronger than the usual triangle inequality and we have

$$d(X,Z) \leq d(X,Y) + d(Y,Z).$$

In summary, then, given any HCS, there corresponds to it a bona fide metric. Conversely, given a distance matrix such as (4.2), the corresponding tree diagram, such as the one in Figure 4.1, can be completely reconstructed, thus yielding the HCS.

Inherent in Johnson's procedure is the definition of distance between two clusters. Furthermore, those two clusters having the minimum distance are clustered. Johnson relates his ideas to the nearest neighbor and furthest neighbor criteria for intercluster distances (definitions 1.8 and 1.9). These two methods yield invariant clustering methods, i.e., the results are invariant under monotone transformations on the similarity matrix.

Jardine and Sibson [179] define a dendogram to be a function which maps the interval $[0,\infty)$ to the set of equivalence relations on P (the set of n objects) such that

(1) every cluster at a given level h' is a union of clusters at level h where $0 \leq h \leq h'$,

(2) for sufficiently large h all objects are in one cluster,

(3) given h, there exists a $\delta > 0$ such that the clusterings at h and $h + \delta$ are exactly alike.

Conditions (1), (2), and (3) are similar to those of Johnson's definition. However, the definition of Jardine and Sibson does not require that all objects are distinct at level $h = 0$. The h in this definition corresponds to α in Johnson's definition. However, Johnson does assume that $\alpha_0 = h = 0$. Jardine and Sibson also discuss the ultrametric inequality and relate their results to various clustering methods. For analagous results see [175], [176], [178], and [180]. These references contain an axiomatic approach to cluster analysis.

Gower and Ross [139] discuss the idea of <u>minimum</u> <u>spanning</u> <u>tree</u> and its relation to single linkage cluster analysis (nearest neighbor, definition 1.8).

<u>Definition 4.1.</u> Given n points in E_p, then a tree spanning these points is any set of straight line segments (edges) joining pairs of points such that

(1)   no closed loops occur,

(2)   each point is visited by at least one line,

(3)   the tree is connected; i.e., any two points are connected by a straight line.

The ideas in this definition come from graph theory (see [154] or [278]).

The length of a tree is the sum of the lengths of the line segments which make up the tree. The <u>minimum</u> <u>spanning</u> <u>tree</u> is then defined to be the tree of minimum length. Gower and Ross discuss two algorithms for finding the minimum spanning tree and present the procedures relevant to these algorithms in another paper [299].

Let $\{d_1, d_2, \ldots, d_h\}$ be the lengths of the edges of a minimum spanning tree, where h is the number of edges. A dendogram can be derived from the set of d_i by grouping those two points which yield the shortest edge and proceeding as in nearest neighbor clustering. Figures 4.2 and 4.3 illustrate the procedure. Clustering by nearest neighbor on the minimum spanning tree is merely clustering on the distance matrix where those distances which are not lengths of the edges of the minimum spanning tree are disregarded. Consequently, clustering on the minimum spanning tree is equivalent to clustering on the distance matrix as was done by Johnson [186]. Zahn [408] discusses minimal spanning trees and their relation to various applications including comments on hierarchical clustering.

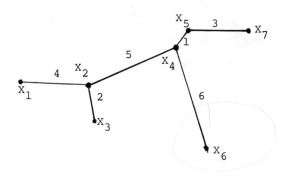

Figure 4.2
Minimum Spanning Tree for Seven Points

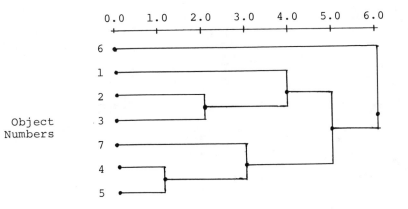

Figure 4.3
Dendogram for the Minimum Spanning Tree of Figure 4.2.

4.2. Comparison of Dendograms or Their Similarity Matrices

It has been seen how in certain cases a distance matrix D has all the information in the corresponding dendogram and vice-versa. For example, the matrix in (4.2) and the dendogram in Figure 4.1. However, this is an ideal situation which does not always exist. Thus it seems desirable to have some objective manner of determining how well a dendogram represents a similarity or distance matrix.

Sokal and Rohlf [335] proposed a measure of correlation between a similarity matrix and values obtained from a dendogram. They also proposed the technique of comparing two dendograms by means of the ordinary correlation coefficient between two sets of values obtained from the two dendograms. This leads to a method of comparing two clustering procedures.

Since a tree or dendogram does not contain all the information present in the similarity matrix we are then confronted with the problem of determining the tree that contains as much information as possible about the similarity matrix. That is, we are faced with the problem of constructing the tree that "best fits" the similarity matrix. Hartigan [162] has presented such a construction. The next two sections contain an exposition of his results.

4.3. Basic Definitions

A <u>tree</u>, denoted by τ, will be considered to be a hierarchical grouping structure. This has been previously described. The term <u>node</u> will be taken to be a single object, a cluster of objects, or a cluster of clusters. Thus, in Figure 4.1, each of the branch points will represent a node.

<u>Definition 4.2</u>. A similarity matrix S is said to have <u>exact</u> <u>tree</u> <u>structure</u> if $s(X_i, X_j) \leq s(X_p, X_q)$ whenever the node into which X_i and X_j are first clustered is at least as high up the tree as the node into which p and q are clustered.

If X_p and X_q are more similar than X_i and X_j then one would want X_p and X_q to occur together in a node (cluster) at an earlier stage in the clustering process than X_i and X_j or equivalently, $s(X_i, X_j) \leq s(X_p, X_q)$ if the smallest cluster containing X_i and X_j also contains X_p and X_q.

Johnson's [186] formulation of tree in section 2.1 involving

the ultrametric inequality satisfies the idea of exact tree structure. Similarly for that of Jardine and Sibson [179].

If a similarity matrix S has exact tree structure then so does
any matrix whose elements are obtained from those of S by means of
a monotonic increasing function. The matrix S requires $n(n-1)/2$
values for its description, however, if S has exact tree structure
it may be described by $2n-1$ real values associated with the nodes
(branch points) of the tree. The value associated with a node is
the value $s(i,j) = s_{ij}$ of any pair (i,j) first clustered into the
node.

Definition 4.3. The distance between two similarity matrices S_1
and S_2 is given by

$$\rho(S_1, S_2) = \sum_{i=1}^{n} \sum_{j=1}^{n} W(i,j)\{s_1(i,j) - s_2(i,j)\}^2/2$$

where $W(i,j)$ is a weight function associated with the similarity
$s(i,j)$.

Definition 4.4. The distance between a similarity matrix S and a
tree τ is given by

$$\rho(S, \tau) = \min_{S*} \ (S, \ S*)$$

where S* is any similarity matrix with exact tree structure τ.

The distance $\rho(S, \tau)$ may be used to determine how well a tree τ represents a similarity matrix S. The primary objective would be to
determine the family of trees $\{\tau_j\}$ with j ranging from $n + 1$ to
$2n - 1$, such that $\rho(S, \tau_j)$ is minimum among trees with j nodes.
Although this is a discrete optimization problem ([292]) with no
known solution for n not very small, it is possible to find trees

which are locally optimal in the sense that no tree in the family
$\{\tau_j\}$ can be improved by performing "local operations" on it that
change it slightly.

4.4. Trees

A <u>tree</u> τ is defined as $\tau = [a_0, A, T]$ where a_0 denotes the root,
A denotes the set (finite) of nodes including a_0, and a mapping T
of A into itself such that for each $k \geq 1$, $T^k a = a$ if and only if
$a = a_0$. For a given dendogram, a_0 may be regarded as the cluster
(node) of all objects n. Given a node b, the transformation T
states which node each node is mapped (grouped) into and may be
regarded as a clustering operation where the nodes in $T^{-1}b$ are
clustered together to form the cluster (node) b. The nodes in $T^{-1}b$
are called the <u>family</u> of b and b is called the <u>parent</u> of the nodes
in $T^{-1}b$. A node b is said to be <u>trivial</u> if $T^{-1}b$ consists of one
node. A node b is said to be <u>barren</u> if $T^{-1}b$ is empty. The set of
barren nodes is denoted by B. The number of barren nodes is denoted
by n(B) and the total number of nodes by n(A). (Actually n(B) = n
and n(A) takes on integer values ranging from n+1 to 2n-1.) The
mapping T which defines the tree begins with the barren nodes and
operates toward the root. Thus, we say that the tree τ is on B.
The ideas here can be compared with those in graph theory by making
reference to [154], [278]. The ideas presented thus far are illustrated in Figure 4.4.

The ideas presented thus far may be related to those of Johnson, and Jardine and Sibson by considering the level at which each
group is created. That is, relate to each node a distance or similarity level α.

In the usual hierarchical clustering scheme the tree does not
have nodes such as 14 and 18 in Figure 4.4. However, one can con-

Similarity

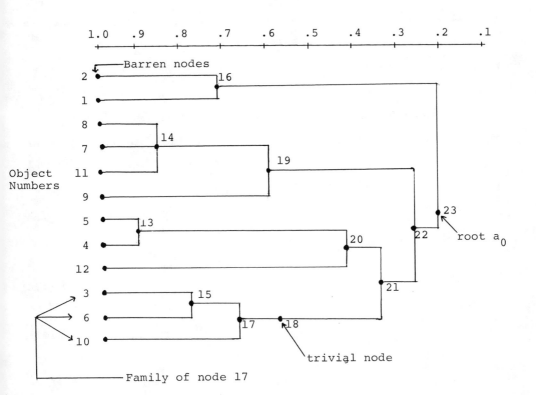

$$B = \{1,2,\ldots,12\}$$

$$a_0 = 23$$

$$T^{-1}(17) = \{3,6,10\}$$

Figure 4.4

sider such nodes as being created by means of "local operations" which are performed to modify a given tree. Such operations are discussed in the next section.

If two nodes a and b are such that for $k \geq 0$, $T^k a = b$ then we say $a \leq b$. The operation $\leq$ defines a partial order on A. If $a, b \in A$, there exists an element, denoted by $c = \sup (a, b)$, such that $a, b \leq c$ and $a, b \leq c^*$ imply $c \leq c^*$; i.e. c is the first node which contains a and b, or $c = T^k a$, $c = T^m b$, $k + m$ minimal. In terms of these ideas a tree τ may be defined as a partial order on A, which has a unique maximal element (a_0), and for which the set $\{b | a \leq b\}$ is linearly ordered for all a.

A <u>similarity tree</u>, denoted by (τ, σ), consists of a tree and a real valued function σ on A, such that

$$\sigma (a) \leq \sigma (b) \quad \text{whenever } b \leq a \tag{4.4}$$

The real value $\sigma(a)$ is called the node similarity of a. Any similarity tree may be represented by a dendogram. In the dendogram, nodes within a family are ordered according to their node similarities or σ-values and ties are broken arbitrarily. This induces a linear ordering on the barren nodes by working from the root. The barren nodes (n-initial nodes) are arranged vertically in the computed order and horizontally according to the values of their node similarities, the largest node similarities falling to the left. The other nodes are arranged vertically in the center of the vertical positions of their families and horizontally according to their node similarities. The complete similarity tree τ may be recovered from the dendogram. In fact, the dendogram is just a visual representation of the abstract notion of tree $\tau = [a_0, A, T]$.

The similarity matrix S is said to have <u>exact</u> <u>tree</u> <u>structure</u>

τ on $\underline{B}$, denoted by S ϵ τ, if for some σ, (τ,σ) is a similarity tree and

$$s(i,j) = \sigma(\sup_{\tau}(i,j)) \qquad (4.5)$$

for all i,j ϵ B, that is, the node similarity of any two nodes i,j ϵ B is the node similarity of the first node into which they are clustered together by T-operations. This is analagous to Johnson's [186] and Jardine and Sibson's [179] definition which associates with nodes i and j the smallest distance level at which i and j occur in the same cluster. If a similarity matrix S has exact tree structure, then the $[n(B)]^2$ (actually $n(n-1)/2$) numbers in S may be recovered from the $n(A)$ numbers in σ.

The distance between two similarity matrices S_1 and S_2 is

$$\rho(S_1,S_2) = \sum_{i=1}^{n} \sum_{j=1}^{n} W(i,j)[s_1(i,j) - s_2(i,j)]^2/2 \qquad (4.6)$$

where W is a symmetric weight function.

The distance between the matrix S and the tree τ is defined by

$$\rho(S,\tau) = \inf_{S^* \epsilon \tau} \rho(S,S^*), \qquad (4.7)$$

where $S^* \epsilon \tau$ means that S^* has exact tree structure. The similarity matrix S^* which minimizes $\rho(S,S^*)$ is called the fit of S to τ. Now, $s^*(i,j)$ is the node similarity of the first node into which i and j are first clustered, i.e. $s^*(i,j) = \sigma(\sup_{\tau}(i,j))$, where σ is some real valued function on A such that $\sigma(a) \geq \sigma(b)$ whenever $a \leq b$. Therefore,

$$\rho(S,\tau) = \inf_{\tau} \sum_{i=1}^{n} \sum_{j=1}^{n} W(i,j)[s(i,j)-\sigma(\sup_{\tau}(i,j))]^2/2$$

where the σ's may be chosen arbitrarily subject to certain inequalities. Determining $\rho(S,\tau)$ is a quadratic programming problem the unique solution of which has been given by Thompson [358]. However, if the only requirement imposed on the minimization of $\rho(S,S^*)$ is that $s^*(i,j) = \sigma(\sup(i,j))$, for σ arbitrary, then

$$\sigma(c) = \sum_{\substack{\sup(i,j) = c \\ \tau}} W(i,j)s(i,j) / \sum_{\substack{\sup(i,j)=c \\ \tau}} W(i,j) \qquad (4.8)$$

If this function happens to satisfy the inequalities $\sigma(a) \geq \sigma(b)$ whenever $a \leq b$, then the corresponding S^* is the fit of S to τ. Actually, the only interest is in trees τ for which τ computed by (4.8) does satisfy the inequalities (see [162]).

The objective of fitting S to τ is to find trees τ with $\rho(S,\tau)$ as small as possible. The concept of distance $\rho(S,\tau)$ allows us to choose "optimum" tree representations for S. The distance $\rho(S,\tau)$ is the mean square error in replacing S by a similarity matrix with exact tree structure τ.

For any tree τ on B, the extension of W, S to A$\times$A is defined by

$$W(a,b) = \sum_{i \leq a, \ j \leq b} W(i,j) \qquad (4.9)$$

$$W(a,b)s(a,b) = \sum_{i \leq a \ j \leq b} W(i,j)s(i,j). \qquad (4.10)$$

The fitted node weights and fitted node similarities are defined by

$$w(c) = W(c,c) - \sum_{Tc'=c} W(c',c') \qquad (4.11)$$

and

$$w(c)\sigma(c) = W(c,c)s(c,c) - \sum_{Tc'=c} W(c',c')s(c',c'), \qquad (4.12)$$

respectively. The σ's defined by equations (4.8) and (4.12) are identical since

$$W(c,c) = \sum_{i \leq c, \, j \leq c} W(i,j) = \sum_{c' \leq c} \sum_{\substack{sup(i,j)=c \\ \tau}} W(i,j)$$

Therefore,

$$w(c) = W(c,c) - \sum_{Tc'=c} W(c',c') = \sum_{\substack{sup(i,j)=c \\ \tau}} W(i,j).$$

Similarly,

$$w(c)\sigma(c) = \sum_{\substack{sup(i,j)=c \\ \tau}} W(i,j)s(i,j).$$

Now, the mean square error $\rho(S,\tau)$ may be written as

$$\rho(S,\tau) = \sum_{i,j \varepsilon B} W(i,j)s^2(i,j) - \sum_{a \varepsilon A} w(a)\sigma^2(a) \qquad (4.13)$$

since

$$\rho(S,\tau) = \sum_{i,j \varepsilon B} W(i,j)[s(i,j) - \sigma(\underset{\tau}{sup}(i,j))]^2$$

$$= \sum_{c \varepsilon A} \sum_{\substack{sup(i,j)=c \\ \tau}} W(i,j)[s(i,j)-\sigma(c)]^2$$

$$= \sum_{c \varepsilon A} \sum_{\substack{sup(i,j)=c \\ \tau}} [W(i,j)s^2(i,j)-W(i,j)\sigma^2(c)]$$

$$= \sum_{i,j \varepsilon B} W(i,j)s^2(i,j) - \sum_{c \varepsilon A} w(c)\sigma^2(c)$$

The equations (4.9)-(4.13) are used in fitting a tree to a similarity matrix S. Given S, the objective is to find trees τ which will minimize $\rho(S,\tau)$, i.e. find the fitted node similarities $\sigma(c)$

which minimize the weighted sum of squares (mean square error)

$$\sum_{c\epsilon A} \sum_{\substack{sup(i,j)=c \\ \tau}} W(i,j)(s(i,j) - \sigma(c))^2 .$$

4.5. Local Operations on Trees

The main objective in fitting τ to S is to find trees τ such that $\rho(S,\tau)$ is a minimum. Now, there always exists a tree τ_{j+1} such that $\rho(S,\tau_{j+1}) \le \rho(S,\tau_j)$. Therefore, the objective will be to find families of trees $\{\tau_j, n(B) < j < 2n(B)+1\}$ such that $\rho(S,\tau_j)$ is a minimum for each j. The only existing method of determining an optimal family is to list all possible trees on B and evaluate $\rho(S,\tau)$ for each tree. This method is highly impractical since the number of trees increases very rapidly with n(B). Thus, one is led to the notion of "locally optimal" family. The procedure is to begin with a family $\{\tau_j\}$ and operate on it iteratively, given a set of "local operations" $\mathcal{L}$. A local operation $L \epsilon \mathcal{L}$ is any operation which changes τ to some other tree, call it $L\tau$. A family of trees $\{\tau_j\}$ is said to be L-optimal, if for every τ_j, $\rho(S,\tau_{j_L}) \le \rho(S,L\tau_j)$, where j_L denotes the number of nodes in $L\tau_j$. This means that the family cannot be improved by use of the operation L. By operating on the trees of a family $\{\tau_j\}$ iteratively, one moves toward an L-optimal family. A family of trees $\{\tau_j\}$ is said to be locally optimal under a set of local operations $\mathcal{L}$, if it is L-optimal for each $L \epsilon \mathcal{L}$. Two desirable properties of local operations are (i) ease of computation and (ii) ρ-reducing ability. Since it is difficult to evaluate an operation in terms of property (ii), except possibly experimentally, property (i) will be used as a guide for evaluation.

It can be shown that if the fitted node similarities given by equation (4.8) do not satisfy the inequality

$$\sigma(a) \geq \sigma(b)$$

whenever $a \leq b$ then a tree can be found with one less node which fits S just as well; i.e., both trees yield the same ρ. This means that the search for locally optimal trees can be restricted to similarity trees. For a justification of this restriction see Hartigan ([162], section 5.).

The local operation L, operating on τ, yields a new tree $L\tau$ with fitted node weights and fitted node similarities denoted by $L\omega$ and $L\sigma$, respectively. The improvement obtained (if any) may be described by the difference $\rho(S,L\tau) - \rho(S,\tau)$. Now, $\rho(S,L\tau)$ may be computed if one knows the $L\omega$ and $L\sigma$ quantities. If $m_1(L)$ denotes the number of ω's and σ's that are changed under the operation L then $m_1(L)$ may be used to indicate how "expensive" it is to compute $\rho(S,L\tau)$.

Once a local operation is carried out the main adjustment is that of modifying some of the values in the extended matrices W,S. All the other quantities associated with the tree must be likewise modified. The procedure is to evaluate a given series of operations on the tree τ and execute that one which reduces ρ the most. The distance $\rho(S,L\tau)$ resulting from the local operation L is compared with $\rho(S,\tau*)$ or $\rho(S,S*)$ where $\tau*$ is the best tree with j_L nodes and S* is the corresponding similarity matrix. One of the computational problems is then to determine the similarity matrix S* with exact tree structure $\tau*$ such that $\rho(S,S*)$ is a minimum.

Following are three local operations that can be performed on a tree:

(i) Rebranch operation: The rebranch operation from a to b denoted by L(a,b), is one which moves a node, a, from the family of Ta and attaches it to the tree at some other node b, $b \nless a$. The function T is then changed to the function LT such that $(LT)a' = Ta'$ for $a' \neq a$ and $(LT)a = b$. The rebranch operation does not change the number of nodes in the tree τ.

(ii) The kill operation $K(a)$ removes the node, a, from the tree τ and transfers its family to Ta. This operation is defined for all nodes except a root or a barren node. In terms of the transformation T, this means that $(KT)a' = Ta'$ for $a' \neq a$, $Ta' \neq a$, and $(KT)a' = Ta$ for $Ta' = a$.

(iii) The make operation $M(a)$ inserts a node $a*$ into the tree τ between a and Ta. If a is the root, then $a*$ becomes the new root. In terms of T, we have $(MT)a' = Ta'$ for $a' \neq a$, $a*$; $(MT)a = a*$; $(MT)a* = Ta$ for $a \neq a_0$; and $(MT)a* = a*$ for $a = a_0$.

The kill and make operations change the number of nodes in the tree τ.

Examples of the tree operations are illustrated in Hartigan's paper. He also gives an example of the application of the preceding procedure on seeking a locally optimal tree and indicates how F-tests may be used to indicate a good size for a tree. Other definitions of exact tree structure and other types of similarity structures besides trees are discussed.

Chapter 5

CLUSTERING BASED ON DENSITY ESTIMATION

5.1. Mode Analysis

One of the clustering techniques discussed in chapter 1,
nearest neighbor, has chaining tendencies; that is, the procedure can
result in elongated clusters. The nearest neighbor technique is not
a variance constraint technique such as those discussed in section
1.6.

In the application of the nearest neighbor technique, due to
chaining tendencies, one might obtain clusters which resemble several
"dense" clusters connected by "sparse" clusters (noise). In the
case of a single variate, the histogram of the data would resemble a
multi-modal distribution. It then seems desirable to utilize a
technique that will detect these modes and assign a cluster to each
mode.

Wishart [396] develops a clustering procedure, entitled mode
analysis for moderate size data sets and outlines its proposed extension for large data sets. The procedure is to first detect
whether the data is multi-modal. In the case of a single variate one
would construct a histogram and temporarily remove the low frequency
(saddle) regions. A cluster can then be associated with each modal
region. Having done this, then the data falling in the saddle region is assigned to their nearest mode. In the case of p variates
the method of first considering a histogram becomes quite awkward.
If each dimension is broken into k classes then one would have to
deal with p^k p-dimensional rectangles. To decide whether a datum
falls in a given class would require p decisions, one for each
dimension. This problem can be circumvented by utilizing spherical
regions. Wishart's one-level algorithm is now stated:

(a) Select a distance threshold r, and a frequency

threshold f.

(b) Compute the similarity matrix S.

(c) Determine the frequency, f_i, of points falling within r units of each data point.

(d) Remove the low frequency points for which $f_i < f$.

(e) Cluster the remaining dense points according to the nearest neighbor technique.

(f) Reallocate each of the points removed in (d) to one of the clusters in (e) according to some criterion. (For example, each non-dense point can be assigned to the cluster with dense point nearest it.)

Wishart further defines a heirarchical algorithm to perform a mode analysis, whereby the user needs to select only the density threshold f. At the first and last cycles of the algorithm only one cluster is defined. At some intermediate cycle the maximum number of clusters is attained. For unimodal data the analysis will reveal only one cluster. See [396] for a detailed description of this heirarchical algorithm and for a fuller discussion.

Zadeh [407] introduced the idea of "fuzzy" set, an idea which can be related to Wishart's mode analysis. According to Zadeh, if E is a space of points then a fuzzy set A in E is characterized by a membership (characteristic) function $f_A(x)$ which associates with each point in E a real number in the interval [0,1], with the value $f_A(x)$ at x representing the "grade of membership" of x in A. The value $f_A(x)$ is analagous to Wishart's frequency threshold value f.

Gitman and Levine [129] present an algorithm which partitions a sample from a multimodal fuzzy set into unimodal fuzzy sets and apply the algorithm to clustering of multivariate data into homogeneous groups. The procedure is similar to Wishart's mode analysis technique.

Since the mode analysis approach to clustering depends very

heavily on the location of modes, it seems that the estimation of the multimodal probability density function relevant to the data would provide a new method of clustering. The remainder of this chapter will discuss this approach.

5.2. Probability Density Function Estimation

Much work has been done in the area of probability density function estimation. It is not the aim here to review density estimation methods. For a brief review of existing methods see Bryan [40]. Bryan proposes a kernel method for estimating a multivariate probability density function and defines a clustering method which is based on density estimation.

The kernel method of estimating a density function is related to the linear integral transformation

$$G(x) = \int K(x - y)\, g(y)\, dy$$

which relates the functions $G(x)$ and $g(y)$. The function $K(x - y)$ is called the kernel of the transformation (see [209]). The kernel method is also called "a weighted average estimate" (see [95], [283], [298]).

The kernel method results in an estimate of the probability density function $f(x)$ of the form

$$\hat{f}(x) = \int K(x - u)\, dF_n(u) = \frac{1}{n} \sum_{j=1}^{n} K(x - x_j)$$

where $K(x - y)$ is the kernel and $F_n(u)$ is the empirical distribution function. Rosenblatt [298] suggests that the kernel function itself be a density function, that is,

$$K(x) \geq 0, \qquad \int K(x)\, dx = 1 .$$

We now review the probability density estimation procedure proposed by Bryan [40].

Let X_1, X_2,..., X_n denote a random sample of size n from some probability density function f(x) which has a nonsingular covariance matrix Σ. A method similar to that of Cacoullos [42] is used to obtain an estimate of the form

$$\hat{f}(x) = \frac{1}{n\alpha^p} \sum_{j=1}^{n} K(\frac{x-x_j}{\alpha})$$

where K is the kernel. The kernel chosen is the multivariate normal probability density function with mean vector 0 and covariance matrix S, that is,

$$K(x) = \frac{1}{(2\pi)^{p/2}|S|^{1/2}} e^{-(1/2)x^T S^{-1} x} \tag{5.1}$$

The estimate is given by

$$\hat{f}(x) = \frac{1}{n\alpha^p (2\pi)^{p/2}|S|^{1/2}} \sum_{j=1}^{n} e^{-\frac{1}{2\alpha^2}(x-x_j)^T S^{-1}(x-x_j)}, \tag{5.2}$$

where X_1, X_2,..., X_n denote observation vectors and

$S = (1/n) \sum_{j=1}^{n} (x_j - \bar{x})(x_j - \bar{x})^T$ is the sample covariance matrix which

is assumed to be nonsingular and this in turn implies that n > p and that S is positive definite.

It is easy to show that $\hat{f}(x)$ in (5.2) is a probability density function, that is,

$$\hat{f}(x) \geq 0 \text{ and } \int \hat{f}(x) dx = 1.$$

The quadratic form, $x^T S^{-1} x$, in the exponent of K(x) in (5.1) is the Mahalanobis distance between x and 0. The matrix I could be used rather than S, thus resulting in Euclidean distance between x and 0, that is $d_2^2(x,0) = x^T x$. The choice between S and I reduces to a choice between Mahalanobis and Euclidean distances.

Euclidean distance is easier and faster to calculate. However, Mahalanobis distance has many advantages. For example, as seen in section 1.3, it is invariant to any nonsingular transformation. This means that $\hat{f}(x)$ is equivalent to $\hat{f}(Ax)$ where A is nonsingular. Thus scaling of the data has no effect on $\hat{f}(x)$. This is not true for $\hat{f}(x)$ when one uses Euclidean distance.

Another property of Mahalanobis distance is that it makes certain grouping criteria equivalent. Three grouping criteria made equivalent under Mahalanobis are (1) tr W, (2) $|T|/|W|$, and (3) tr $W^{-1}B$, where T, B, and W are, respectively the total, between, and within scatter matrices discussed in section 1.4. Criteria (2) and (3) have been suggested by Friedman and Rubin [122] and all three are discussed by them. The clustering procedure to be dis-cussed in the next section can be used to obtain an optimal grouping, at least locally, with respect to criteria (2) and (3). A technique discussed in section 1.5 obtains an optimal grouping with respect to criterion (1).

A problem inherent in using the estimate $\hat{f}(x)$ in (5.2) is the choice of the proper value for α. The choice of α is of great importance and a poor choice will yield unsatisfactory estimates.

The choice of α is based on the information theoretical inequality

$$r = E\{\ln \frac{f(x)}{\hat{f}(x)}\} = \int f(x) \ln \frac{f(x)}{\hat{f}(x)} \, dx \geq 0 \qquad (5.3)$$

Kullback [217] proves that one has equality in (5.3) if and only if $f(x) = \hat{f}(x)$ for almost all x. The procedure is to choose that α that minimizes r. The result (5.3) can be written as

$$m = E[\ln \hat{f}(x)] = \int f(x) \ln \hat{f}(x) \, dx \leq \int f(x) \ln f(x) \, dx = E[\ln f(x)]$$

with equality if and only if $f(x) = \hat{f}(x)$ for almost all x. Thus the minimization of r can be looked at as a problem of maximizing m. The procedure in selecting α is to maximize an estimate of m rather than m itself. If $\hat{f}$ is based on $x_1, x_2, \ldots, x_n$ and if $y_1, y_2, \ldots, y_k$

denotes another sample of size k then an estimate of m is given by

$$\hat{m} = \frac{1}{k} \sum_{i=1}^{k} \ln \hat{f}(y_i).$$

If a second sample if not available then we would want an estimate of $\hat{f}$ based on x_1, x_2, ..., x_n. An estimate of the form

$$\hat{m} = \frac{1}{k} \sum_{i=1}^{k} \ln \hat{f}(x_i) \qquad (5.4)$$

is biased and could lead to negative values of α. The bias in the estimate (5.4) can be reduced by jackknifing [81], [141]. Let $\hat{h}_j(x)$ denote the estimate of $f(x)$ when x_j is omitted. Thus

$$\hat{h}_j(x) = \frac{1}{(n-1)\alpha^p} \sum_{\substack{i=1 \\ i \neq j}}^{n} K\left(\frac{x-x_i}{\alpha}\right),$$

and $\hat{m}$ is given by

$$\hat{m} = \frac{1}{n} \cdot \sum_{j=1}^{n} \ln \hat{h}_j(x_j).$$

The derivative of $\hat{m}$ is, after simplification

$$\frac{d\hat{m}}{d\alpha} = \frac{1}{n\alpha^3} \sum_{j=1}^{n} \frac{\displaystyle\sum_{\substack{i \neq j}}^{n} D^2(x_i,x_j) e^{-\frac{1}{2\alpha^2} D^2(x_i,x_j)}}{\displaystyle\sum_{\substack{i \neq j}}^{n} e^{-\frac{1}{2\alpha^2} D^2(x_i,x_j)}} - \frac{p}{\alpha},$$

where

$$D^2(x_i,x_j) = (x_i - x_j)^T S^{-1} (x_i - x_j).$$

The choice of α is then the solution of $\frac{d\hat{m}}{d\alpha} = 0$. A technique such as the Newton-Raphson can be used to solve this equation. For a discussion of such techniques see chapter 3 in Isaacson and Keller [172].

5.3. Clustering Based on Density Estimation

It has been noted that one of the advantages of using the density estimate $\hat{f}(x)$ given by (5.1) is that it makes the three

grouping criteria (1) tr W, (2) $|T|/|W|$, and (3) tr $W^{-1}B$ equivalent.
The three criteria are made equivalent by first adjusting the observation vectors y_1, y_2, ..., y_n according to $x_i = y_i - \bar{y}$,
$i = 1, 2, ..., n$ so that they have mean vector 0 and then trans-
forming the x vectors according to CX where C is nonsingular and
$CTC^T = I$. A theorem in [40] shows that Euclidean distance in the
transformed space is proportional to Mahalanobis distance in the
original space. Thus a clustering algorithm can be used to group
data with respect to the three criteria without having to transform
the data if Mahalanobis distance is used in the original space.

Most of the clustering techniques discussed in the previous
chapters have been formulated on heuristic or intuitive grounds and
are of a deterministic nature. The technique based on density
function estimation is a statistical approach and leads to a well
(or at least better)-defined notion of cluster. This statistical
approach is based on locating the modes. Two options are available;
(1) the modes can be estimated directly from the data or (2) the
multivariate probability density function f(x) can be estimated by
$\hat{f}(x)$ after which the modes of $\hat{f}(x)$ are computed.

In the heuristic approach the choice of clustering technique
depends largely on the data. In the statistical approach a cluster
is defined in terms of the characteristics of the density function
from which the data is derived. Thus in this approach one can state
more precisely what the goal of clustering is, use statistical estimation to objectively approach that goal, and use statistical testing
to evaluate how close to the goal the results are.

The statistical approach to clustering proposed in [40] consists
of first estimating the multivariate density function that the data
are taken from. Then a "hill climbing" technique is used to assign
samples to clusters which are defined in terms of the modes of $\hat{f}(x)$.
This hill climbing procedure can be described as follows.

The estimate $\hat{f}(x)$ is first obtained. Then the closest sample (nearest neighbor) and second closest sample (second nearest neighbor) to each sample x_i are determined. A sample becomes part of a "path" or "hill" of $\hat{f}(x)$ if its nearest neighbor yields a larger value for $\hat{f}$. When all the samples have been examined several paths are formed which climb the "hills" (or hill) of $\hat{f}(x)$. The samples in each path are grouped to form the clusters.

A path will terminate with a sample whose nearest neighbor has a smaller $\hat{f}(x)$. Basing the algorithm strictly on nearest neighbor can result in too many paths (clusters). To alleviate this situation each sample's second nearest neighbor is also considered. If a sample x_i has a nearest neighbor x_k and a second nearest neighbor x_j such that $\hat{f}(x_k) \leq \hat{f}(x_i) \leq \hat{f}(x_j)$ then the path is continued. This two nearest-neighbor procedure is used for the initial clustering.

The above approach will always leave two disconnected paths on both sides of a "hill" (mode). To remedy this situation $\hat{f}(x)$ can be compared at three points between each peak and its nearest peak. The values of $\hat{f}(x)$ at these three points will indicate whether two nearest peaks form a "hill" in wnich case the samples corresponding to the two paths are grouped, or whether a "valley" lies between them in which case the two paths are left disconnected. The procedure is repeated until all samples are associated with one hill or until a valley exists between each peak and its nearest peak. The three points between two nearest peaks can be chosen, say, on a line one fourth, one half, and three fourths of the way between peaks.

Bryan [40] describes computer programs that performs the density estimation and the clustering described above. He gives examples for which the technique gives satisfactory results. One of his examples utilizes the well-known Fisher's Iris data and is illustrated in the next chapter.

5.4 Remarks

From sections 5.2 and 5.3 it is evident that Bryan's procedure could be modified in a variety of ways. For example, a different type of density estimation could be used. His estimate, however, does have a certain flexibility in terms of the quantity α. Some of the aspects of his clustering procedure are quite arbitrary and could be modified, however, this is a minor point.

Another definition of a cluster, assuming a particular probability model, is proposed by Ling [231]. He defines a cluster and two indices for measuring compactness and relative isolation and develops a probability theory relevant to the sampling distributions of the two indices.

Wolfe [402] describes a clustering method based on the decomposition of a mixture of multivariate distributions and provides complete documentation and listing of an IBM 360/65 program which implements the method.

Koontz and Fukunaga [208] present a general expression for a nonparametric clustering criterion which is analagous to (4.6). They describe a clustering procedure, based on their criterion, which is valley-seeking, that is it detects the low frequency (saddle) regions.

Chapter 6

APPLICATIONS

6.1 Application to Remote Sensing Data

In the remote sensing problem [167], [226], [227], one is faced
with an image (or scene) which is a rectangular region with r rows
(scan lines) and c columns (numbers of resolution elements per scan
line) of one resolution element (an individual). Each cell (individual, I) generates a p x 1 measurement vector X_{ij}, i = 1,2,...,r,
and j = 1,2,...,c. In order to recognize a scene, one must perform
rc = n discriminate tasks "as effectively as possible", (if the
scene is classified point by point).

The Manned Spacecraft Center acquired Ball and Hall's [15], [16]
[18], ISODATA (Iterative Self-Organizing Data Analysis Technique)
program as a prospective tool for clustering remote sensing data
(multispectral scanner data). The object of the clustering process
is two-fold [194],(a) to check the homogeneity of the multispectral
scanner data, that is, to see if it is necessary to split a training
class of resolution elements into several unimodal training subclasses, and (b) to cluster the data of a whole flight line so that
complete classification of these data reduces to a mere relabelling
of the various groups.

Ball and Hall's [15], [16], [18], iterative procedure has been
briefly discussed in section 1.6. The procedure first forms k clusters by selecting k objects at random and then assigning each of the
remaining n-k objects to that center which is nearest it. The
cluster centroids are computed and any two clusters I and J are
combined if D_{IJ}^2 is less than a threshold r. A cluster is split if
the variance s_x^2 within the cluster of any one variate x exceeds a
threshold s^2. The variance s_I^2 of each resultant cluster I is thus

constrained by $s_I^2 \leq p\ s^2$, where p is the number of variates. The cluster centroids replace the original cluster centers and the process is continued until it stabilizes (convergence is attained). A description of the ISODATA program is contained in Holley [166]. The original ISODATA program used Euclidean distance. The one described in [166] uses a weighted Euclidean distance.

Kan and Holley [194] propose a final recommended version of ISODATA. They replace Euclidean distance with the ℓ_1 distance in assigning objects to their nearest reference points. The measure of variation in each dimension s_x^2 remains the same. The weighted Euclidean distance between two clusters is used for D_{IJ}^2.

An application of Kan and Holley's recommended version of ISODATA to multispectral scanner data will now be given. The example is taken from Kan and Holley [194]. The data consists of r = 35 scan lines and c = 45 columns or sample points on each scan line. Thus the data consists of n = 35(45) = 1575 observations. There were four channels sampled on the multispectral scanner, that is, p = 4. The splitting threshold was 4.5 and the combining threshold was 3.2. This means that on a particular iteration a cluster was split along the jth dimension if the variance along the jth dimension exceeded 4.5. On the other hand if two clusters are closer than 3.2 units then they are combined into one cluster.

The program took 12 iterations yielding seven clusters with 565, 132, 219, 201, 180, 224, and 54 members. Table 6.1 gives the summary statistics for each cluster and Table 6.2 gives the matrix of intercluster distances at the end of 12 iterations. The data for Kan and Holley's analysis was taken from [409]. Further detail regarding this example is given in [194].

TABLE 6.1

Summary Statistics for Each Cluster

Cluster No.		Channel 1	2	3	4
1	mean	182.59	176.91	187.33	200.39
	stand. dev.	1.779	2.671	1.814	1.862
2	mean	178.35	172.44	173.98	187.72
	stand. dev.	3.619	2.840	2.532	2.720
3	mean	166.68	163.63	167.10	181.35
	stand. dev.	2.250	2.150	2.614	2.541
4	mean	179.42	174.99	181.01	194.69
	stand. dev.	2.308	2.259	2.337	2.130
5	mean	179.77	170.24	157.64	166.74
	stand. dev.	2.200	1.728	2.134	2.348
6	mean	163.10	158.60	159.57	174.53
	stand. dev.	2.362	2.645	3.356	2.946
7	mean	181.89	173.52	164.56	173.74
	stand. dev.	2.738	1.940	3.306	4.006

TABLE 6.2

Table of Intercluster Distances

Cluster #	1	2	3	4	5	6	7
1	0.0	8.7	16.0	4.5	22.3	19.7	13.6
2	8.7	0.0	6.5	4.2	10.9	9.9	5.5
3	16.0	6.5	0.0	11.1	9.9	4.4	8.2
4	4.5	4.2	11.1	0.0	16.5	14.7	9.4
5	22.3	10.9	9.9	16.5	0.0	9.6	4.0
6	19.7	9.9	4.4	14.7	9.6	0.0	10.0
7	13.6	5.2	8.2	9.4	4.0	10.0	0.0

6.2. Application of Density Estimation Technique to Fisher's Iris Data [40]

Fisher's Iris data consists of four characteristics ($p = 4$) for three species of Iris. There are fifty samples corresponding to each species, that is, $n = 150$. The three species are Iris setosa, Iris versicolor, and Iris virginica, and the four characteristics are sepal length, sepal width, petal length and petal width. The cluster analysis, using Mahalanobis distance, yielded nine clusters. Table 6.3 gives the make-up of the clusters according to species.

TABLE 6.3

Number of samples belonging to each species-cluster combination

	Cluster #								
Species	1	2	3	4	5	6	7	8	9
Iris setosa	49	1	0	0	0	0	0	0	0
Iris versicolor	0	14	13	17	3	3	0	0	0
Iris virginica	0	7	5	6	18	0	3	5	6

If one assigns cluster 1 to Iris setosa, clusters 2, 3, 4, and 6 to Iris versicolor and clusters 5, 7, 8, and 9 to Iris virginica then there are 22 misclassifications. Repeating the above analysis using Euclidean distance rather than Mahalanobis distance yielded only 8 misclassifications. However, in this case there resulted 17 clusters as compared to 9 in the previous case.

For the sake of comparison, Zahn's [408] graph-theoretic method when applied to the same data (with one duplicate point omitted) yielded 9 clusters and 45 misclassifications.

Chapter 7

HISTORICAL COMMENTS

In recent years there has been immense interest in the area of numerical classification and in particular in cluster analysis. A great majority of the publications dealing with cluster analysis have occurred in a wide variety of journals which treat a diversity of topics. Consequently, a work which unifies the various results and presents them in a coherent fashion is lacking. The first six chapters (with the possible exception of chapter 2) have been an attempt to do this.

In this chapter we present some brief remarks pertinent to the development of cluster analysis over the past four decades.

The initial description and statement of what is now known as cluster analysis were formulated by Tryon [361] in 1939. Tryon and Bailey ([371], 1970) in their recent book discuss their BC TRY computer system for performing cluster and factor analysis from the point of view of the behaviorial or social scientist. A book by Fisher ([108], 1968) discusses particular methods relevant to the aggregation problem in economics. Cole ([57], 1969) presents a collection of papers presented at a colloquium in numerical taxonomy. The recent book by Jardine and Sibson [180] presents a mathematical treatment of methods related to biological taxonomy but which have wider application. Their treatment is also contained in [175], [177], [178], [179]. They present an axiomatic approach to cluster analysis. Anderberg ([5], in press) has written a book in cluster analysis, the aim of which is to present a unified self-contained treatment of cluster analysis at an elementary level, and discusses various background topics pertinent to cluster analysis. The book of Sokal and Sneath ([336], 1963) is a good reference for persons working in biology oriented areas however, their book is inadequate for contemporary researchers.

Ball [13] presents an excellent review and comparison of some "cluster seeking" techniques. He categorizes such techniques as falling into seven classes, namely, (1) probabilistic, (2) signal detection, (3) clustering, (4) clumping, (5) eigenvalue, (6) minimal mode seeking, and (7) miscellaneous. The most widely used, and what is generally meant by cluster analysis, are (3) clustering and (4) clumping techniques. The hierarchical techniques discussed in chapter 1 are clumping techniques and the "minimum variance constraint" techniques mentioned in chapter 1 are clustering techniques. The minimal mode seeking technique requires categorization information for its operation. Hall's eigenvalue technique categories are akin to factor analytic and principal component techniques in multivariate analysis.

Another category that could be constructed is that comprising density estimation techniques discussed in chapter 5. Hall's "probabilistic techniques" category could be extended to cover these techniques.

The "clustering" techniques are generally regarded as the most efficient and easily interpreted techniques. However, in taxonomic studies "clumping" techniques are the most popular. Historically, clumping techniques were the first clustering methods used due to their use in numerical taxonomy. Sokal and Sneath [336] is a good reference for these clumping methods.

A clustering technique that has found wide acceptance is Ball and Hall's [15], [16], [18] ISODATA (Iterative Self-Organizing Data Analysis Technique). This technique was mentioned briefly in chapter 1 as a "minimum variance constraint" method. It is being used in grouping remote sensing data (multispectral scanner data) at the National Aeronautics and Space Administration, Manned Spacecraft Center. Its use there has been documented [166], [191], [193], and a final recommended version of ISODATA is given by Kan and Holley [194]. An application of ISODATA to remote sensing data (multi-spectral scanner data) is given in the previous chapter.

Other useful techniques are Hartigan's adding algorithm (Lecture Notes) which is also described in Kan and Holley [194]. This is a divisive hierarchical clustering technique. Kan and Holley also discuss other methods which have been implemented and could be useful computational-wise. They include [325], [208], and [404], and [302].

In the area of pattern recognition the object is, as in cluster analysis to sort the data into groups. However, in pattern recognition the category information about each observation is known. Nagy [270] gives an excellent survey of pattern recognition. His paper contains 148 references.

Cluster methods could be classified into two broad categories, (1) divisive and (2) agglomerative. One must not confuse algorithm with method. A given method could be implemented by means of a divisive or agglomerative algorithm. Divisive methods work by successive partitioning of a set of objects and agglomerative methods work by successive pooling of subsets (clusters) of objects. Divisive methods have been proposed by Edwards and Cavalli-Sforza [93], Macnaughton-Smith et. al. [235], and Rescigno and Maccacaro [293]. Agglomerative techniques are more abundant, some of which have been described in chapter 1, [223], [234], [387]. The hierarchical clustering algorithm given in chapter 1 is an agglomerative algorithm. Agglomerative procedures are not in general iterative in that they require a specific procedure that determines when two clusters are combined.

Some methods are, strictly speaking, neither divisive nor agglomerative such as, for example, the methods in [18], [33], and [237] discussed in chapter 1. Another method is the dynamic programming method such as that of Jensen [183], discussed in chapter 4.

Very little has been done in the way of evaluating clustering techniques. It is very difficult to determine whether one method is "better" than another method. Which method one uses depends on the purpose of the study and on the type of data at hand. Consequently, it may not be realistic to compare two given methods. A comparison of the methods of Sokal and Michener [334], Edwards and Cavalli-Sforza [93], and Williams and Lambert [394] are given by Gower [137]. Rand [288] presents objective criteria for comparing two different clusterings on the same set of data. For other comments on the performance of cluster methods see Jardine and Sibson ([180], chapter 2), Borko [34], and Green and Rao [144].

Fisher and Van Ness [102] define several admissibility conditions which would be desirable under almost any circumstances and compare various standard clustering methods with them. These admissibility conditions are formulated so as to reject procedures which would yield "bad" clusterings. However, an admissibility condition, although rejecting procedures which yield bad clusterings, might reject, in some instances, procedures which yield reasonable clusterings. Some of the admissibility conditions in [102] are also discussed in Hartigan [162], Johnson [186], and Jardine and Sibson [178]. The study by Fisher and Van Ness [103], which discusses admissible discriminant analysis, is also of interest here.

The number of clusters that one wishes to obtain from a cluster method may not be known subsequent to the analysis. Rather, one may wish to determine the number m from the results. Hierarchical (agglomerative) and divisive methods allow the analyst to

determine m from considering the various hierarchic levels. The
dynamic programming approach requires the number m to be known
prior to the analysis. Pattern recognition techniques also require
m to be known.

Another term that is quite vague in working with cluster analysis is the idea of "convergence" of a method. This term was used
in discussing some of the "variance constraint" methods in chapter
1. The clusters resulting from a method should be unique. If the
data is multimodal with well defined modes then the clusters will
tend to be unique. However, if the data is "flat" such as that
arising from a uniform distribution then the clusters need not be
unique. This just means there are no well-defined clusters or
merely a single cluster (m = 1).

The concepts discussed in this monograph have been introduced
as techniques for clustering objects or individuals. These methods
however could be employed in clustering variates or characteristics.
Hartigan [161] presents a technique for two-way clustering, that
is, the clustering of individuals and variates simultaneously.

Cluster Analysis is related to other multivariate analysis
techniques such as principal component analysis, discriminant
analysis, and factor analysis.

Discriminant analysis can be used to obtain a preliminary
partition of the data. This partition can then be updated by
iteratively reassigning the observations by recomputation of the
discriminant function until the "best" partition, as dictated by
the discriminant analysis, is obtained. Casetti [45] and Hung
and Dubes [170] provide programs that implement this technique.
Mayer [261] proposes a similar method which uses a single variable
of primary importance. In a later paper Mayer [263] gives a
method where the data matrix is first reduced in terms of p' < p

principal components after which a distance criteria and the linear discriminant function are used to generate the clusters. Urbakh [375] proposes a method of dividing a heterogeneous multi-variate population into groups by using discriminant analysis to extrapolate successively until a best division, in terms of mis-classification probability, is obtained.

REFERENCES

[1] Abraham, C. T., A note on a measure of similarity used in
     the DICO experiment, Appendix I, Quarterly Report 3,
     Vol. 1, Contract AF 19(626)-10.

[2] Abraham, C. T., Evaluation of clusters on the basis of random
     graph theory, Yorktown Heights, N.Y.: IBM Corporation,
     IBM Res. Memo, November, 1962.

[3] Adhikari, B. P. and Joshi, D. D., Distance discrimination et
     resume exhaustif, Pbls. Inst. Statist., Vol. 5, (1956),
     57-74.

[4] Aitken, M. A., The correlation between variate values and ranks
     in a doubly truncated normal distribution, Biometrika,
     Vol. 53, Parts 1/2, (1966), 281-282.

[5] Anderberg, M. R., Cluster analysis for applications, (in press),
     December, 1971.

[6] Anderberg, M. R., An Annotated Bibliography of Cluster Analysis,
     Mechanical Engineering Department, University of Texas at
     Austin (in preparation), (1972).

[7] Anderson, T. W., An Introduction to Multivariate Statistical
     Analysis, John Wiley & Sons, Inc., New York, (1958).

[8] Archer, W. B., Computation of Group Job Descriptions from
     Occupational Survey Data, Report Number PRL-TR-66-12,
     Personnel Research Laboratory, Lackland AFB, Texas, 31 pp.

[9] Armstrong, J. S. and Soelberg, P., On the interpretation of
     factor analysis, Psychol. Bull., 70 (1968), 361-364.

[10] Astrahan, M. M., Speech Analysis by Clustering, or the Hyper-
     phoneme Method, Stanford Artificial Intelligence Project
     Memo AIM-124, Stanford University, 22 pp., (1970).

[11] Balas, E., An additive algorithm for solving linear programs
     with zero-one variables, Operations Res., 13 (1965),
     517-546.

[12] Balinski, M. L., Integer programming: methods, uses and
     computation, Management Sci., 12 (Nov. 1965), 253-313.

[13] Ball, G. H., A Comparison of Some Cluster-Seeking Techniques,
     Report Number RADC-TR-66-514, Stanford Research Inst.
     Menlo Park, California, 47 pp., (1966).

[14] Ball, G. H., Classification Analysis, Technical Note, Stanford
     Research Inst. Menlo Park, California, (1970).

[15]   Ball, G. H., Data Analysis in the social sciences--what about
       details?, American Federation of Information Processing
       Societies Conference Proceedings: 1965 Fall Joint Computer
       Conference, 27 (1965), Part 1, 533-560, (Washington:
       Spartan Books; London: Macmillan).

[16]   Ball, G. H. and Hall D. J., A clustering technique for summar-
       izing multivariate data, Behavioral Sciences, Vol. 12,
       No. 2, (March, 1967), 153-155.

[17]   Ball, G. H. and Hall, D. J., Background information on
       clustering techniques, Stanford Research Inst., (July,
       1968).

[18]   Ball, G. H. and Hall, D. J., ISODATA, A Novel Method of
       Data Analysis and Pattern Classification, Technical
       Report, Menlo Park, California: Stanford Research
       Inst., 72 pp., (1965).

[19]   Ball, G. H. and Hall, D. J., PROMENADE--An On-Line Pattern
       Recognition System, Report Number RADC-TR-67-310,
       Stanford Research Inst., 124 pp., (1967).

[20]   Baker, F. B., Latent class analysis as an association model
       for information retrieval, in Stevens, Giuliano and
       Heilprin (eds.), Statistical Association Methods for
       Mechanized Documentation, National Bureau of Standards
       Miscellaneous Publication Number 269, U.S. Government
       Printing Office, Washington, D.C., (1965), 149-155.

[21]   Bartels, P. H., Bahr, G. F., Calhoun, D. W. and Wied, G. L.,
       Cell recognition by neighborhood grouping technique in
       TICAS, Acta Cytologica, Vol. 14, No. 6, (1970), 313-324.

[22]   Barton, D. E. and David, F. N., Spearman's 'Rho' and the
       matching problem, Brit. J. Statist. Psychol., 9 (1956),
       69-73.

[23]   Bass, B. M., Iterative inverse factor analysis--a rapid method
       for clustering persons, Psychometrika, Vol. 22, No. 1,
       (March, 1957), 105-107.

[24]   Baxendale, P., An empirical model for computer indexing,
       Machine Indexings Progress and Problems, American Univer-
       sity, Washington, D.C., (Feb. 13-17, 1961), 267.

[25]   Beale, E. M. L., Euclidean cluster analysis, Bull. I.S.I.,
       43, 2(1969), 92-94.

[26]   Beale, E. M. L., Selecting an optimum subset, in J. Abadie
       (ed.), Integer and Nonlinear Programming, Amsterdam:
       North Holland Publishing Company, (1970).

[27]   Bellman, R. E. and Dreyfus, S. E., Applied Dynamic Programming,
       Princeton, N.J.: Princeton University Press, (1962).

[28]   Benders, J. F., Partitioning procedures for solving mixed-
       variables programming problems, Numerische Mathematik,
       4, (Feb. 1962), 238-252.

[29] Bhattacharyya, A., On a measure of divergence between two statistical populations defined by their probability distributions, Bull. Calcutta Math. Soc., Vol. 35, (1943), 99-109.

[30] Birnbaum, A. and Maxwell, A. E., Classification procedures based on Bayes' formula, in L. J. Cronbach and Goldine C. Gleser (eds.), Psychological tests and personnel decisions, Urbana: University of Illinois Press, (1965).

[31] Bledsoe, W. W., A corridor-projection method for determining orthogonal hyperplanes for pattern recognition, unpublished report, Panoramic Research Cor., Palo Alto, California (1963).

[32] Block, H. D., Knight, B. W. and Rosenblatt, F., The perception: A model for brain functioning, II, Rev. Modern Phys., Vol. 34, No. 1, (Jan. 1962), 135-142.

[33] Bonner, R. E., On some clustering techniques, IBM Journal, 22, (Jan. 1964), 22-32.

[34] Borko, H., Blankenship, D. A. and Burket, R. C., On-Line Information Retrieval Using Associative Indexing, RADC-TR-68-100, Systems Development Corporation, (1968), 124 pp.

[35] Bottenberg, R. A. and Christal, R. E., An Iterative Technique for Clustering Criteria Which Retains Optimum Predictive Efficiency. WADD-TN-61-30, Lackland AFB, Texas: Personnal Research Laboratory, Wright Air Development Division, (March, 1961).

[36] Boulton, D. M. and Wallace, C. S., A program for numerical classification, Comput. J., Vol. 13, (1970), 63-69.

[37] Bradley, R. A., Katti, S. K. and Coons, I. J., Optimal scaling for ordered categories, Psychometrika, Vol. 27, No. 4, (1962), 355-374.

[38] Brennan, E. J., An analysis of the adaptive filter, General Electric Elec. Lab. Tech. Information Series Report R61 ELS-20, Syracuse, N.Y., (1961).

[39] Bryan, J. G., Calibration of Qualitative or Quantitative Variables for Use in Multiple-Group Discriminant Analysis, The Travelers Weather Research Center, Hartford, Conn., 26 pp.

[40] Bryan, J. K., Classification and clustering using density estimation, Ph.D. Dissertation, University of Missouri, Columbia, Missouri, (Aug., 1971).

[41] Butler, G. A., A vector field approach to cluster analysis, Pattern Recognition, 1 (1969), 291-299.

[42] Cacoullos, T., Estimation of multivariate density, Ann. Inst.
      Statist. Math., Vol. 18, (1966), 179-189.

[43] Campbell, J. P., A hierarchical cluster analysis of the core
      courses in an engineering curriculum, J. Exp. Educ.,
      (1966), 35, 63-69.

[44] Carroll, J. B., The nature of the data, or how to choose a
      correlation coefficient, Psychometrika, Vol. 26, No. 4,
      (1961), 347-372.

[45] Casetti, E., Classificatory and Regional Analysis by Dis-
      criminant Iterations, TR-12, Contract Nonr-1228(26),
      Northwestern University, Evanston, Ill., 99 pp., (1964).

[46] Castellan, N. J., Jr., On the estimation of the tetrachoric
      correlation coefficient, Psychometrika, Vol. 31, No. 1,
      (1966), 67-73.

[47] Cattell, R. B., A note on correlation clusters and cluster
      search methods, Psychometrika, (Sept., 1944), 9, 169-184.

[48] Cattell, R. B., Factor analysis: An introduction to
      essentials II. The role of factor analysis in research,
      Biometrics, Vol. 21, No. 2, (1965), 405-435.

[49] Cattell, R. B. and Coulter, M. A., Principles of behavioural
      taxonomy and the mathematical basis of the taxonomic
      computer program, Brit. J. Math. Statist, Psychol.,
      19 (1966), 237-269.

[50] Charnes, A. A. and Cooper, W. W., Management Models and
      Industrial Applications of Linear Programming, Vol. 1,
      New York: John Wiley & Sons, Inc., (1961).

[51] Chernoff, H., Metric Considerations in Cluster Analysis,
      Technical Report No. 67, Department of Statistics,
      Stanford University, Stanford, California, 16 pp., (1970).

[52] Christal, R. E. and Ward, J. H., Jr., Applications of a new
      clustering technique which minimizes loss in terms of
      any criterion specified by the investigator, paper pre-
      sented at the meeting of the American Psychological
      Association, New York City, (Sept., 1961).

[53] Christal, R. E. and Ward, J. H., Jr., The MAXOF clustering
      model. Lackland AFB, Texas: Personnel Research Divi-
      sion Air Force Human Resources Laboratory (AFSC),
      (1970) in press.

[54] Christal, R. E. and Ward, J. H., Jr., Use of an objective
      function in clustering people or things into mutually
      exclusive categories, paper presented at Conference
      on Cluster Analysis of Multivariate Data, New Orleans,
      La., (Dec., 1966).

[55] Clark, P. J., An extension of the coefficient of divergence for use with multiple characters, Copeia 2 (1952), 61-64.

[56] Cochran, W. G. and Hopkins, C. E., Some classification problems with multivariate qualitative data, Biometrics, Vol. 17, No. 1, (1961), 10-32.

[57] Cole, A. J., Numerical Taxonomy, Academic Press, New York, (1969).

[58] Cole, L. C., The measurement of interspecific association, Ecology 30 (1949), 411-424.

[59] Cole, L. C., The measurement of partial interspecific association, Ecology 38 (1957), 226-233.

[60] Constantinescu, P., A method of cluster analysis, Brit. J. Math. Statist. Psychol. 20(1), (1967), 93-106.

[61] Cooper D. B., Nonsupervised adaptive signal detection and pattern recognition, Raytheon Report, (Oct. 22, 1963).

[62] Cooper, D. B. and Cooper, P. W., Adaptive pattern recognition and signal detection without supervision, IEEE International Convention Record, Part 1, (1964), 246-256.

[63] Cooper, D. B. and Cooper, P. W., Nonsupervised adaptive signal detection and pattern recognition, Information and Control, Vol. 7, No. 3, (Sept., 1964).

[64] Cooper, W. W. and Majone, G., A description and some suggested extensions for methods of cluster analysis, Internal Working Memorandum, Carnegie-Mellon University.

[65] Cover, T. M. and Hart, P. E., Nearest-neighbor pattern classification, IEEE Trans. Inf. Theory, 13, (1967), 21-27.

[66] Cox, D. R., Note on grouping, J. Amer. Statist. Assoc., 52, (1957), 543- 547.

[67] Cramer, H., On the composition of elementary errors, Skand. Aktuarietids, Vol. 11, (1928), 13-74 and 141-180.

[68] Cramer, H., The Elements of Probability Theory and Some of its Applications, John Wiley & Sons, Inc., New York, (1946).

[69] Crawford, R. M. M. and Wishart, D., A rapid classification and ordination method and its application to vegetation mapping, J. Ecology, Vol. 56, No. 2, (1968), 385-404.

[70] Crawford, R. M. M. and Wishart, D., A rapid multivariate method for the detection and classification of groups of ecologically related species, J. Ecology, Vol. 55, No. 2, (1967), 505-524.

[71] Cronbach, L. J. and Gleser, G. C., Assessing the similarity between profiles, Psychol. Bull., Vol. 50, No. 6, (1953), 456-473.

[72] Dagnelie, P., On different methods of numerical classification, Rev. Statist. Appl. 14, III (1966), 55-75.

[73] Daly, R. F., Adaptive binary detection, Stanford Elec. Lab. Tech. Report No. 2003-2, Stanford, California, (June 26, 1961).

[74] Daly, R. F., The adaptive binary-detection problem on the real line, Stanford Elec. Lab. Report SEL-62-030, Stanford, California, (Feb., 1962).

[75] Daniels, H. H., Rank correlation and population models, J. Roy. Statist. Soc., B, 12 (1950), 171-181.

[76] Darling, D. A., The Kolmogorov-Smirnov, Cramer-Von Mises tests, Ann. Math. Statist., Vol. 28, (Dec. 1957), 823-838.

[77] Day, N. E., Estimating the components of a mixture of normal distributions, Biometrika, Vol. 56, (1969), 463-474.

[78] David, F. N., Barton, D. E., Ganeshalingham, S., Harter, H. L., Kim, P. J., Merrington, M. and Walley, D., Normal Centroids, Medians and Scores for Original Data, Cambridge University Press, Cambridge, England, (1968).

[79] David, S. T., Kendall, M. G. and Stuart, A., Some questions of distribution in the theory of rank correlation, Biometrika 38 (1951), 131-140.

[80] Demiremen, F., Multivariate Procedures and FORTRAN IV Program for Evaluation and Improvement of Classifications, Computer Contribution 31, State Geological Survey, The University of Kansas, Lawrence, 51 pp., (1969).

[81] Dempster, A. P., Elements of Continuous Multivariate Analysis, Reading, Massachusetts: Addison-Wesley Publishing Co., (1969).

[82] Dice, L. R., Measures of the amount of ecological association between species, Ecology 26 (1945), 297-302.

[83] Dobes, J., Partitioning algorithms, Inf. Processing Math., 13, (1967), 307-313.

[84] Doyle, L. B., Breaking the Cost Barrier in Automatic Classification, Professional Paper SP-2516, System Development Corp., Santa Monica, California, 62 pp.,(1966).

[85] Dubes, R. C., Information Compression, Structure Analysis and Decision Making with a Correlation Matrix, Michigan State University, East Lansing, Michigan, 243 pp., (1970).

[86] Dubin, R., Typology of Empirical Attributes: Multi-dimensional Typology Analysis (MTA), TR-5, University of California at Irvine, 17 pp., (1971).

[87] Dubin, R. and Champoux, J. E., Typology of Empirical Attributes: Dissimilarity Linkage Analysis (DLA), Technical Report 3, University of California at Irvine, 31 pp., (1970).

[88] Duncan, D. B., Multiple range and multiple F tests, Biometrics, Vol. 11, No. 1, (1955), 1-42.

[89] Durbin, J. and Stuart, A., Inversions and rank correlation coefficients, J. Roy. Statist. Soc., B, 13 (1951), 303-309.

[90] Eades, D. C., The inappropriateness of the correlation coefficient as a measure of taxonomic resemblance, Systematic Zoology, Vol. 14, No. 2, (1965), 98-100.

[91] Eddy, R. P., Class Membership Criteria and Pattern Recognition, Report 2524, Naval Ship Research and Development Center, Washington, D. C., 47 pp., (1968).

[92] Edwards, A. W. F., The measure of association in a 2x2 table, J. Roy Statist. Soc., Series A, Vol. 126, Part 1, (1963), 109-114.

[93] Edwards, A. W. F. and Cavalli-Sforza, L. L., A method for cluster analysis, Biometrics, Vol. 21, No. 2, (1965), 362-375.

[94] Eisen, M., Elementary Combinatorial Analysis, Gordon and Breach Science Publishers, (1969).

[95] Elkins, T. A., Cubical and spherical estimation of multivariate probability density, J. Amer. Statist. Assoc., Vol. 63, No. 324, (1968), 1495-1513.

[96] Engleman, L. and Hartigan, J. A., Percentage points of a test for clusters, J. Amer. Statist. Assoc., Vol. 64, (1969), 1947-1948.

[97] Ericson, W. A., A note an partitioning for maximum between sum of squares, Appendix C in J. A. Sonquist and J. N. Morgan, The Detection of Interaction Effects, Survey Research Center, Institute for Social Research, University of Michigan, Ann Arbor.

[98] Estabrook, G. F., A mathematical model in graph theory for biological classification, J. Theoret. Biol., 12 (1966), 297-310.

[99] Eusebio, J. W. and Ball, G. H., ISODATA-LINES - A program for describing multivariate data by piecewise-linear curves, Proceedings of International Conference on Systems Science and Cybernetics, University of Hawaii, Honolulu, Hawaii, (Jan. 1968), 560-563.

[100] Farris, J. S., On the Cophenetic Correlation coefficients, Systematic Zoology, Vol. 18, No. 3, (1969), 279-285.

[101] Firschein, O., Fischler, M., Automatic subclass determination for pattern recognition applications, Trans. PGEC, EC-12, No. 2 (April, 1963).

[102] Fisher, L. and Van Ness, J., Admissible clustering procedures,
       Biometrika 58, (1971), 91-104.

[103] Fisher, L. and Van Ness, J., Admissible discriminant analysis,
       J. Amer. Statist. Assoc. 68 (1973), 603-607.

[104] Fisher, R. A., The precision of discriminant functions, Ann.
       Eugenics, Vol. 10, (1940), 422-429.

[105] Fisher, R. A., Statistical Methods for Research Workers,
       Hafner Publishing Co., New York, Thirteenth Edition,
       reprint, (1963).

[106] Fisher, R. A., The use of multiple measurements in taxonomic
       problems, Ann. Eugenics, Vol. VII, Part II, (1936),
       179-188.

[107] Fisher, R. A. and Yates, F., Statistical Tables for Biological,
       Agricultural and Medical Research, Oliver and Boyd, London,
       (1953).

[108] Fisher, W. D., Clustering and Aggregation in Economics, The
       Johns Hopkins Press, Baltimore, Maryland, (1968).

[109] Fisher, W. D., On a pooling problem from the statistical
       decision viewpoint, Econometrika, (1953), 21, 567-585.

[110] Fisher, W. D., On grouping for maximum homogeneity, J. Amer.
       Statist. Assoc., Vol. 53, (1958), 789-798.

[111] Fisher, W. D., Simplification of economic models, Econometrika,
       Vol. 34, No. 3, (July, 1966), 563-584.

[112] Flake, R. H. and Turner, B. L., Numerical classification for
       taxonomic problems, J. Theoret. Biol., Vol. 20 (1968),
       260-270.

[113] Forgy, E. W., Classification so as to relate to outside
       variables, in M. Lorr and S. B. Lyerly (eds.), Final
       Report, Conference on Cluster Analysis of Multivariate
       Data, Catholic University of America, Washington, D. C.,
       (1966), pp. 13.01-13.12.

[114] Forgy, E. W., Cluster Analysis of Multivariate Data: Effici-
       ency Versus Interpretability of Classifications, paper
       presented at Biometric Society meetings. Riverside,
       California, (abstract in Biometrics, Vol. 21, No. 3,)
       (1965), p. 768.

[115] Forgy, E. W., Detecting 'Natural' Clusters of individuals,
       Western Psychological Association Meetings, Santa Monica,
       California, (Apr. 19, 1963).

[116] Forgy, Edward W., Detecting 'Natural' clusters of individuals,
       report at Western Psychological Association, Dept, of
       Psychiatry, University of California Medical Center,
       Los Angeles, California, (April, 1963), 1-10.

[117] Forgy, E. W., Evaluation of several methods for detecting sample mixtures from different N-dimensional populations, American Psychology Association Meetings, Los Angeles, Calif., (Sept. 9, 1964). (Available from author at Center for Health Sciences, U.C.L.A., Los Angeles, Calif.).

[118] Fortier, J. J., Contributions to item selection, Tech. Rep. no. 2, Laboratory for Quantitative Research in Education, Stanford University, (1962).

[119] Fortier, J. and Solomon, H., Clustering procedures, Multivariate Analysis, ed. by P. R. Krishnaiah, Academic Press, N.Y., (1966), 493-506.

[120] Fralick, S. C., The synthesis of machines which learn without a teacher, IEEE Trans. on Information Theory, Vol. IT-13, (1967), 57-64.

[121] Fralick, S. C., The synthesis of machines which learn without a teacher, Tech. Report No. 6103-8, Stanford University, (April, 1964).

[122] Friedman, H. P. and Rubin, J., On some invariant criteria for grouping data, J. Amer. Statist. Assoc., Vol. 62, (1967), 1159-1178.

[123] Froemel, E. C., A comparison of computer routines for the calculation of the tetrachoric correlation coefficient, Psychometrika, Vol. 36, No. 2, (1971), 165-174.

[124] Fu, K. S., Langrebe, D. A. and Phillips, T. L., Information processing of remotely sensed agricultural data, Proc. IEEE, Vol. 57, No. 4, (April, 1969), 639-653.

[125] Fukunaga, K. and Koontz, W. L., A criterion and an algorithm for grouping data, IEEE Trans. On Computers, Vol. C-19, (Oct. 1970), 917-923.

[126] Garfinkel, R. S. and Nemhauser, G. L., Optimal political districting by implicit enumeration techniques, Management Sci., Vol. 16, No. 8, (1970), B495-B508.

[127] Garfinkel, R. and Nemhauser, G. L., The set-partitioning problem: Set covering with equality constraints, Operations Res., 17, (Sept. - Oct. 1969), 848-856.

[128] Gengerelli, J. A., A method for detecting subgroups in a population and specifying their membership, J. Psychology, Vol. 55, (1963), 457-468.

[129] Gitman, I. and Levine, M. D., An algorithm for detecting unimodal fuzzy sets and its application as a clustering technique, IEEE Trans. on Comput., Vol. C-19, No. 7, (1970), 583-593.

[130] Glaser, E. M., Signal detection by adaptive filters, IRE Trans. on Info. Theory, Vol. IT-7, No. 2, (April, 1961).

[131] Gleason, A. M., A search problem in the n-cube, Proc. Symposium in Appl. Math. Amer. Math. Soc., 10, (1960), 175-178.

[132] Glover, F., A multiphase-dual algorithm for the zero-one integer programming algorithm, Operations Res., Vol. 13, No. 6, (Nov. - Dec. 1965), 879-919.

[133] Goldberger, A. S., Impact Multipliers and Dynamic Properties of the Klien-Goldberger Model, Amsterdam: North-Holland Publishing Co., (1959).

[134] Gomory, R. E., All-integer integer programming algorithm, in J. L. Muth and G. L. Thompson (eds.), Industrial Scheduling, Englewood Cliffs, N. J.: Prentice-Hall (1963).

[135] Goodman. L. A. and Kruskal, W. H., Measures of association for cross classifications, J. Amer. Statist. Assoc., Vol. 49, (1954), 732-764.

[136] Goodman, L. A. and Kruskal, W. H., Measure of association for cross-classification, II, J. Amer. Statist. Assoc., Vol. 54, (1959), 123-163.

[137] Gower, J. C., A comparison of some methods of cluster analysis, Biometrika, Vol. 23, No. 4, (1967), 623-637.

[138] Gower, J. C., Some distance properties of latent root and vector methods used in multivariate analysis, Biometrika, Vol. 53, No. 3/4, (1966), 325-338.

[139] Gower, J. C. and Ross, G. J. S., Minimum spanning trees and single linkage cluster analysis, Appl. Statist., Vol. 18, No. 1, (1969), 54-64.

[140] Grason, J., Methods for the Computer-Implemented Solution of a Class of "Floor-Plan" Design Problems, Ph.D. Dissertation, Electrical Engineering Department, Carnegie-Mellon University, Pittsburgh, Pennsylvania, 374 pp. (1970).

[141] Gray, H. L. and Schucany, W. R., The Generalized Jackknife Statistic, New York: Marcel Dekker, Inc., (1972).

[142] Green, P. E. and Carmone, F. J., Multi-dimensional Scaling and Related Techniques in Marketing Analysis, Allyn and Bacon Inc., Boston, (1970).

[143] Green, P. E., Frank, R. E. and Robinson, P. J., Cluster analysis in test market selection, Management Sci., 13, (1967), 13-387-400.

[144] Green, P. E. and Rao, V. R., A note on proximity measures and cluster analysis, J. Marketing Research, Vol. VI, (1969), 359-364.

[145] Hadley, G., Nonlinear and dynamic programming, Addison-Wesley, Reading, Massachusetts, (1964).

[146] Haggard, E. A., Intra-Class Correlation and the Analysis
     of Variance, Dryden, N.Y., (1958).

[147] Hall, A. V., Avoiding informational distortions in automatic
     grouping programs, Systematic Zoology, Vol. 18, No. 3,
     (1969), 318-329.

[148] Hall, D. J., Ball, G. H., Wolf, D. E. and Eusebio, J.,
     PROMENADE: An Improved Interactive-Graphics Man/Machine
     System for Pattern Recognition, Report Number RADC-TR-
     68-572, Stanford Research Institute, Menlo Park, Calif.
     173 pp., (1969).

[149] Hamdan, M. A., Estimation of class boundaries in fitting a
     normal distribution to a qualitative multinomial distri-
     bution, Biometrics, Vol. 27, No. 2, (1971), 457-459.

[150] Hamdan, M. A., On the polychoric method for estimation of
     [Rho] in contingency tables, Psychometrika, Vol. 36,
     No. 3, (1971), 253-259.

[151] Hammer, P. L. and Rudeanu, S., Boolean Methods in Operations
     Research and Related Areas, New York: Springer-Verlag,
     (1968).

[152] Hanson, N. R., Patterns of Discovery: An Inquiry into the
     Conceptual Foundations of Science, Cambridge University
     Press, New York, (1958).

[153] Haralick, R. M. and Dinstein, I., An iterative clustering
     procedure, IEEE Trans. on Systems, Man and Cybernetics,
     Vol. SMC-1, No. 3, (July, 1971), 275-289.

[154] Harary, F., Graph Theory, Addison-Wesley, Reading, Massachu-
     setts, (1969).

[155] Harding, E. F., The number of partitions of a set of N
     points in k dimensions induced by hyperplanes, Proc.
     Edinburg Math. Soc., Vol. 15, (1967), 285-289.

[156] Harding, E. F., The probabilities of rooted tree-shapes
     generated by random bifurcation, Advances in Appl.
     Probability, Vol. 3, No. 1, (1971), 44-77.

[157] Harman, H. H., Modern Factor Analysis, University of
     Chicago Press, Chicago, Illinois, (1960).

[158] Harrison, J., Cluster Analysis, Metra, 7, (1968), 513-518.

[159] Harter, H. L., Expected values of normal order statistics,
     Biometrika, Vol. 48, Part 1/2, (1961), 151-165.

[160] Hartigan, J. A., Clustering a Data Matrix, working paper,
     Department of Statistics, Yale University, New Haven,
     Connecticut, (1970), 55 pp.

[161] Hartigan, J. A., Direct clustering of a data matrix, J.
     Amer. Statist. Assoc., Vol. 67, (1972), 123-129.

[162] Hartigan, J. A., Representation of similarity matrices by trees, J. Amer. Statist. Assoc., Vol. 62, (1967), 1140-1158.

[163] Hartigan, J. A., Using subsample values as typical values, J. Amer. Statist. Assoc., Vol. 64, (1969), 1303-1317.

[164] Hasselblad, V., Estimation of parameters for a mixture of normal distributions, Technometrics, Vol. 8, (1966), 431-446.

[165] Henschke, C. I., Manpower Systems and Classification Theory, Ph.D. Dissertation, University of Georgia, Athens, Georgia, 261 pp., (1969), cited in Dissertation Abstracts, Vol. 30, No. 8, p. 3911-B.

[166] Holley, W. A., Description and user's guide for the IBM 360/44 ISODATA PROGRAM, Lockheed Electronics Co., Inc., HASD, Houston, Texas, Tech. Rep. 640-TR-030, (Sept., 1971).

[167] Holmes, R. A., and MacDonald, R. B., "The physical basis of systems design for remote sensing in agriculture", Proceedings IEEE, Vol. 57, April 1969, pp. 629-639.

[168] Holzinger, K. J. and Harman, H. H., Factor Analysis, Chicago Press, Chicago, Ill., (1941).

[169] Huang, F., Per field classifier for agriculture applications, LARS Information Note 060569, Purdue University, Lafayette, Indiana, (June, 1969).

[170] Hung, A. Y. and Dubes, R. C., An Introduction to Multiclass Pattern Recognition in Unstructured Situations, Interim Scientific Report No. 12, Division of Engineering Research, Michigan State University, East Lansing, Michigan, 66 pp., (1970).

[171] Hyvarinen, L., Classification of Qualitative Data, Brit. Info. Theory J., (1962), 83-89.

[172] Isaacson, E. and Keller, H. B., Analysis of Numerical Methods, John Wiley and Sons, Inc., New York, (1966).

[173] Jaccard, P., Nouvelles Recherches sur la distribution florale, Bull. Soc. Vand. Sci. Nat. 44, (1908), 223-270.

[174] Jancey, R. C., Multidimensional group analysis, Australian J. Botany, Vol. 14, No. 1, (1966), 127-130.

[175] Jardine, C. J., Jardine, N. and Sibson, C., The structure and construction of taxonomic hierarchies, Mathematical Biosciences, Vol. 1, No. 2, (1967), 173-179.

[176] Jardine, N., Algorithm, methods, and models, in the simplification of complex data, Comput. J., 13, (1970), 116-117.

[177] Jardine, N., Towards a general theory of clustering, <u>Bio-metrics</u>, 25, (1969), 609-610.

[178] Jardine, N. and Sibson, R., The construction of hierarchic and nonhierarchic classifications, <u>Comput. J.</u>, Vol. 11, (1968) 177-184.

[179] Jardine, N. and Sibson, R., A model for taxonomy, <u>Math. Biosci.</u>, 2, (1968), 465-482.

[180] Jardine, N. and Sibson, R., Mathematical Taxonomy, John Wiley and Sons, New York, (1971).

[181] Jeffreys, H., Theory of probability, Oxford University Press, (1948).

[182] Jeffreys, H., An invariant for the prior probability in estimation problems, <u>Proc. Roy. Soc. A.</u>, Vol. 186, (1946), 454-461.

[183] Jensen, R. E., A dynamic programming algorithm for cluster analysis, <u>Operations Res.</u>, 12, (Nov. - Dec. 1969), 1034-1057.

[184] John, S., On identifying the population of origin of each observation in a mixture of observations from two normal populations, <u>Technometrics</u>, Vol. 12, (1970), 553-565.

[185] Johnson, P. O., The quantification of qualitative data in discriminant analysis, <u>J. Amer. Statist. Assoc.</u>, Vol. 45, (1950), 65-76.

[186] Johnson, S. C., Hierarchical clustering schemes, <u>Psychometrika</u>, Vol. 32, No. 3, (Sept. 1967), 241-254.

[187] Kahl, J. A. and Davis, J. A., A comparison fo indexes of socio-economic status, <u>Amer. Sociol. Rev.</u>, Vol. XXI, 3 (1956).

[188] Kailath, T., The divergence and Bhattacharyya distance measures in signal selection, <u>IEEE Trans. on Comm. Tech.</u>, Vol. COM-15, (Feb. 1967), 52-60.

[189] Kaminuma, T., Takekawa, T. and Watanabe, S., Reduction of clustering problem to pattern recognition, <u>Pattern Recognition</u>, Vol. 1, (1969), 195-205.

[190] Kan, E. P. F., Data clustering: An overview, Lockheed Electronics Co., Inc., HASD, Houston, Texas, Tech. Rep. 640-TR-080, (March, 1972).

[191] Kan, E. P. F., ISODATA: Thresholds for splitting clusters, Lockheed Electronics Co., Inc., HASD, Houston, Texas, Tech. Rep. 640-TR-058, (January, 1972).

[192] Kan, E. P. F., On an iterative clustering technique, Lockheed Electronics Co., Inc., HASD, Houston, Texas, Tech. Memo. #LEC TM642-214, (November, 1971).

[193]   Kan, E. P. F. and Holley, W. A., Experience with ISODATA,
        Lockheed Electronics Co., Inc., HASD, Houston, Texas,
        Tech. Memo TM 642-354, (March, 1972).

[194]   Kan, E. P. F. and Holley, W. A., More on clustering techniques
        with final recommendations on ISODATA, Lockheed Electro-
        nics Co., Inc., HASD, Houston, Texas, Tech. Rep. #LEC
        640-TR-112, (May, 1972).

[195]   Kaskey G. et al, Cluster formation and diagnostic signifi-
        cance in psychiatric symptom evaluation, Proc. Fall Jt.
        Computer Conf., (1962), p. 285.

[196]   Kazmierezak, H. and Steinbuch, K., Adaptive systems in pattern
        recognition, IEEE Trans. on Electronic Computers, Vol.
        EC-12, No. 6, (December, 1963).

[197]   Keifer, J. and Wolfowitz, J., Consistency of the maximum
        likelihood estimator in the presence of infinitely
        many incidental parameters, Ann. Math. Statist.,
        Vol. 27, (1956), 887-906.

[198]   Kendall, M. G., A Course in Multivariate Analysis, Hafner
        Publishing Company, New York, Fourth Impression, (1968).

[199]   Kendall, M. G., A new measure of rank correlation, Biometrika,
        30 (1938), 81-93.

[200]   Kendall, M. G., Discrimination and classification, Multi-
        variate Analysis, ed. by P. R. Krishnaiah, Academic
        Press, N.Y., (1966), 165-184.

[201]   Kendall, M. G., Rank Correlation Methods, Griffin, London,
        2nd edition, (1965).

[202]   Kendall, M. G., Kendall, S. F. H. and Smith, B. B., The
        distribution of Spearman's coefficient of rank correla-
        tion in a universe in which all rankings occur an equal
        number of times, Biometrika 30 (1938), 251-273.

[203]   Kendall, M. G. and Stuart, A., The Advanced Theory of Sta-
        tistics, Vol. II, Inference and Relationship, Charles
        Griffin & Co., Ltd., London, (1961).

[204]   King, B., Stepwise clustering procedures, J. Amer. Statist.
        Assoc., (1967), 62, 86-101.

[205]   Kochen, M., Techniques for information retrieval research:
        State of the art, presented at IBM World Trade Corpora-
        tion Information Retrieval Symposium at Blaricum, Holland,
        (Nov., 1962), to be published in the proceedings of the
        symposium.

[206]   Kochen, M. and Wong, E., Concerning the possibility of a
        cooperative information exchange, IBM Journal of Re-
        search and Development, Vol. 6, No. 2, (April, 1962),
        270-271.

[207] Kolmogorov, A. N., Sulla determinazionne empirica di une
       legge di distribuzione, Giorn, dell'Instit. degli att.,
       Vol. 4, (1933), 83-91.

[208] Koontz, W. L. and Fukunaga, K., A nonparametric valley-
       seeking technique for clustering analysis, IEEE Trans.
       on Computers, Vol. C-21, No. 2, (Feb., 1972), 171-178.

[209] Korn, G. A. and Korn, T. M., Mathematical Handbook for
       Scientists and Engineers, New York: McGraw-Hill Book
       Company, (1968).

[210] Kraft, C. H., Some conditions for consistency and uniform
       consistency of statistical procedures, University of
       California Publications in Statistics, (1955).

[211] Kruskal, J. B., Multidimensional scaling by optimizing
       goodness-of-fit to a nonmetric hypothesis, Psychometrika,
       29 (1964), 1-27.

[212] Kruskal, J. B., Nonmetric multidimensional scaling: A numeri-
       cal method, Psychometrika, 29, No. 2, (June, 1964),
       115-129.

[213] Kruskal, J. B., Jr., On the shortest spanning subtree of a
       graph and the traveling salesman problem, Proc. Amer.
       Math. Soc., No. 7, (1956), 48-50.

[214] Kruskal, W. H., Ordinal measures of association, J. Amer.
       Statist. Assoc., Vol. 53, (1958), 814-861.

[215] Kshirsagar, A. M., Goodness of fit of an assigned set of
       scores for the analysis of association in a contingency
       table, Ann. Inst. Statist. Math., Vol. 22, No. 2, (1970),
       295-306.

[216] Kullback, S. and Leibler, R. A., On information and suf-
       ficiency, Ann. Math. Statist., Vol. 22, (1951), 79-86.

[217] Kullback, S., Information Theory and Statistics, New
       York, Dover Publications, Inc., (1968).

[218] Labovitz, S., In defense of assigning numbers to ranks,
       American Sociological Review, Vol. 36, No. 3, (1971),
       520-521.

[219] Labovitz, S., Some observations on measurement and statistics,
       Social Forces, Vol. 46, No. 2, (1967), 151-160.

[220] Lancaster, H. O. and Hamdan, M. A., Estimation of the correla-
       tion coefficient in contingency tables with possibly non-
       metrical characters, Psychometrika, Vol. 29, No. 4, (1964),
       383-391.

[221] Lance, G. N. and Williams, W. T., Computer programs for hier-
       archical polythetic classification ('Similarity Analyses'),
       Comput. J., Vol. 9, No. 1, (1966), 60-64.

[222]   Lance, G. N. and Williams, W. T., Computer program for mono-
        thetic classification ('Association Analysis'), Comput.
        J., Vol. 8, No. 3, (1965), 246-249.

[223]   Lance, G. N. and Williams, W. T., A general theory of classi-
        ficatory sorting strategies. 1. Hierarchical systems,
        Comput. J., Vol. 9, No. 4, (1967), 373-380.

[224]   Lance, G. N. and Williams, W. T., A general theory of
        classificatory sorting strategies II. Clustering systems,
        Comput. J., Vol. 10, No. 3, (1967), 271-276.

[225]   Lance, G. N. and Williams, W. T., A generalized sorting
        strategy for computer classifications, Nature, Vol. 212,
        (1966), p. 218.

[226]   Landgrebe, D. A. and LARS Staff, "LARSYAA, A Processing
        system for airborne earth resource data", LARS Infor-
        mation Note 091968, Purdue University, Lafayette,
        Indiana, September 1969.

[227]   Landgrebe, D. A. and Phillips, T. L., "A multichannel image
        data handling system for agriculture remote sensing",
        Proc. Seminar on Computerized Image Handling Tech-
        niques, Washington, D. C., June, 1967, pp. XIT-1 to 10.

[228]   Lemke, C. E. and Spielberg, K., Direct search algorithms
        for zero-one and mixed integer programming, Operations
        Res., Vol. 15, No. 5, (Sept. - Oct., 1967), 892-914.

[229]   Levine, M. D., Feature Selection: a survey, Proc. IEEE,
        Vol. 57, No. 8, August 1969, 1391-1408.

[230]   Lewis, P. M., The characteristic selection problem in
        recognition systems, IRE Transaction on Information
        Theory, IT-8, February 1962, 171-178.

[231]   Ling, R. F., A probability theory of cluster analysis,
        J. Amer. Statist. Assoc., 68, (1973), 159-164.

[232]   Litofsky, B., Utility of Automatic Classification Systems
        for Information Storage and Retrieval, Ph.D. Dissertation,
        University of Pennsylvania, cited in Dissertation Ab-
        stracts, Vol. 30, No. 7, (Jan., 1970), 3264-B.

[233]   MacNaughton-Smith, P., The classification of individuals
        by the possession of attributes associated with a
        criterion, Biometrics, 19, (1963), 364-366.

[234]   MacNaughton-Smith, P., Some statistical and other numerical
        techniques for classifying individuals. (home office
        res. rpt. no. 6) H.M.S.O., London, (1965).

[235]   MacNaughton-Smith, P. and Williams, W. T., Dale, M. B. and
        Mockett, L. G., Dissimilarity analysis: A new technique
        of hierarchical division, Nature, Vol. 201, (1964),
        p. 426.

[236] MacQueen, J. B., Some methods for classification and analysis of multivariate observations, Proc. of the Fifth Berkeley Symposium on Mathematical Statistics and Probability, Vol. 1, (1967), 281-297.

[237] MacQueen, J. B., Some methods for classification and analysis of multivariate observations, Western Management Sci. Inst., University of California, working paper 96, (1966).

[238] McCammon, R. B. and Wenninger, G., The Dendograph, Computer Contribution 48, State Geological Survey, The University of Kansas, Lawrence, (1970), 28 pp.

[239] McCammon, R. B., The dendograph: A new tool for correlation, Geological Society of America Bulletin, Vol. 79, (1968), 1163-1670.

[240] McQuitty, L. L., Agreement analysis: Classifying persons by predominant patterns of responses, Brit. J. Statist. Psychol., Vol. 9, (1956), p. 5.

[241] McQuitty, L. L., Single and multiple hierarchical classification by reciprocal pairs and rank order types, Educational and Psychological Measurement,26, (1966), 253-265.

[242] McQuitty, L. L., Agreement analysis: Classifying persons by predominant patterns of response, Brit. J. Statist. Psychol.,9, (1956), 5-16.

[243] McQuitty, L. L., Capabilities and improvements of linkage analysis as a clustering method, Educational and Psychological Measurement, 24, (1964), 441-456.

[244] McQuitty, L. L., Elementary linkage analysis for isolating orthogonal and oblique types and typal relevances, Educational and Psychological Measurement, 17, (1957), 207-229.

[245] McQuitty, L. L., Hierarchical syndrome analysis, Educational and Psychological Measurement 20, (1960), 293-304.

[246] McQuitty, L. L., Improved hierarchical syndrome analysis of discrete and continuous data, Educational and Psychological Measurement 26, (1966), 577-582.

[247] McQuitty, L. L., Multiple hierarchical classification of institutions and persons with reference to union--measurement relations and psychological well-being, Educational and Psychological Measurement, 22, (1962), 513-531.

[248] McQuitty, L. L., Rank order typal analysis, Educational and Psychological Measurement, 23, (1963), 55-61.

[249] McQuitty, L. L., Typal Analysis, Educational and Psychological Measurement, 21, (1961), 677-696.

[250] McRae, D. J., MIKCA: A FORTRAN IV Iterative K-means cluster analysis program, Behavioral Science, Vol. 16, No. 4, (1971), 423-424.

[251] Mahalonobis, P. C., Analysis of race mixture in Bengal, J. Asiat. Soc. (India), Vol. 23, (1925), 301-310.

[252] Mahalanobis, P. C., On the generalized distance in statistics, Proc. Natl. Inst. Sci. (India), Vol. 12, (1936), 49-55.

[253] Majone, G., Distance-based cluster analysis and measurement scales, working paper no. 17, University of British Columbia, Vancouver, B.C., Canada, (Nov., 1968).

[254] Majone, G. and Sanday, P. R., On the Numerical Classification of Nominal Data, Report Number RR-118, Graduate School of Industrial Administration, Carnegie-Mellon University, (1968).

[255] Marill, T. and Green, D. M., On the effectiveness of receptors in recognition systems, IEEE Trans. Information Theory, IT-9, January 1963.

[256] Marriot, F. H. C., A problem of optimum stratification, Biometrics, Vol. 26, (1970), 845-847.

[257] Marriot, F. H. C., Practical problems in a method of cluster analysis, Biometrics, Vol. 27, (1971), 501-514.

[258] Mattson, R. L. and Damman, J. E., A technique for determining and coding subclasses in pattern recognition problems, IBM Journal, (July, 1965), 294-302.

[259] Matusita, K., On the theory of statistical decision functions, Ann. Instit. Statist. Math. (Tokyo), Vol. 3, (1951), 17-35.

[260] Mayer, L. S., Comment on the assignment of numbers to rank order categories, American Sociological Review, Vol. 35, No. 5, (1970), 916-917.

[261] Mayer, L. S., A method of cluster analysis when there exist multiple indicators of a theoretic concept, Biometrics, Vol. 27, No. 1, (1971), 143-155.

[262] Mayer, L. S., A note on treating original data as interval data, American Sociological Review, Vol. 36, No. 3, (1971), 518-519.

[263] Mayer, L. S., A method of cluster analysis, paper presented at the joint statistical meetings, Fort Collins, Colorado, August 23-26, 1971.

[264] Medgyessy, Pal, Decomposition of Super-positions of Distribution Functions, Publishing House of Hungarian Academy of Science, Budapest, (June, 1961).

[265] Michener, C. D. and Sokal, R. R., A quantitative approach to a problem in classification, Evolution, Vol. 11, (June, 1957), 130-162.

[266] Michener, C. D. and Sokal, R. R., A quantification of systematic relationships and phylogenetic trends, Proc. Xth International Congress of Entomogy I, (1957), 409-415.

[267] Morishima, H. and Oka, H., The pattern of interspecific variations in the genus oryza: Its quantitative representation by statistical methods, Evolution, 14, (1960), 153-165.

[268] Morrison, D. G., Measurement problems in cluster analysis, Management Sci., 13, (1967), 13-775-780.

[269] Morrison, D. F., Multivariate Statistical Methods, McGraw-Hill Book Company, New York, (1967)

[270] Nagy, G., State of the art of pattern recognition, Proc. IEEE, Vol. 56, (1968), 836-862.

[271] Nagy, G. and Tolaba, J., Nonsupervised crop classification through airborne MSS observations, IBM J. Res. and Dev., (March, 1972).

[272] Needham, R. M., A method for using computers in information classification, Proc. I.F.I.P. Congress, 62, (1962), p. 284.

[273] Needham, R. M., The theory of clumps, II, Report M. L., 139, Cambridge Language Research Unit, Cambridge, Eng., (March, 1961).

[274] Needham, R. M. and Jones, K. S., Keywords and clumps, J. Documentation, Vol. 20, (1964), p. 5.

[275] Nunnally, J., The analysis of profile data, Psychol. Bull., Vol. 59, No. 4, (1962), 311-319.

[276] Okajima, M., Stark, L., Whipple, G. and Yasui, S., Computer pattern recognition techniques: Some results with real electrocardiographic data, IEEE Trans. on Bio-Medical Electronics, Vol. BME-10, No. 3, (July, 1963).

[277] Olds, E. J., The 5% significance levels of sums of squares of rank differences and a correction, Ann. Math. Statist. 20, (1949).

[278] Ore, O., Theory of graphs, Amer. Math. Soc., Providence, R. I., (1962).

[279] Orr, D. B., A new method for clustering jobs, J. Appl. Psychol. (1960), 44, 44-59.

[280] Parker-Rhodes, A. F., Contributions to the theory of clumps, I.M.L. 138, Cambridge Language Research Unit, Cambridge England, (March, 1961).

[281] Parks, J. M., Classification of mixed mode data by r-mode factor analysis and q-mode cluster analysis on distance functions, in A. J. Cole (ed.), Numerical Taxonomy, Academic Press, New York, (1969), 216-223.

[282] Parks, J. M., <u>FORTRAN IV Program for Q-Mode Cluster Analysis</u> <u>on Distance Function with Printed Dendogram</u>, Computer Contribution 46, State Geological Survey, The University of Kansas, Lawrence, (1970).

[283] Parzen, E., On estimation of a probability density function and mode, <u>Ann. Math. Statist.</u>, Vol. 33, (1962), 1065-1076.

[284] Patrick, E. A. and Hancock, J. C., The non-supervised learning of probability spaces and recognition of patterns, Tech. Report, Purdue University, Lafayette, Ind., (1965).

[285] Patrick, E. A. and Shen, L. Y. L., Interactive use of problem knowledge for clustering and decision making, <u>IEEE</u> <u>Trans. Computers</u>, Vol. C-20, (Feb., 1971), 216-223.

[286] Pearson, W. H., Estimation of a correlation measure from an uncertainty measure, <u>Psychometrika</u>, Vol. 31, No. 3, (1966), 421-433.

[287] Pearson, E. S. and Hartley, H. O., <u>Biometric Tables for</u> <u>Statisticians</u>, Cambridge University Press, Cambridge, England, (1954).

[288] Rand, W. M., The Development of Objective Criteria for Evaluating Clustering Methods, Ph.D. Dissertation, UCLA, 138 pp. Cited in <u>Dissertation Abstracts</u>, Vol. 30, No. 11, (May, 1970), 4932-B.

[289] Rao, C. R., The use and interpretation of principal components analysis in applied research, <u>Sankhya</u>, Series A, Vol. 26, (1964), 329-358.

[290] Rao, C. R., The utilization of multiple measurements in problems of biological classification, <u>J. Roy. Statist.</u> <u>Soc.</u>, Series B, 10, (1948), 159-203.

[291] Rao, M. R., Cluster Analysis and Mathematical Programming, Journal of the American Statistical Association, <u>J.</u> <u>Amer. Statist. Assoc.</u>, Vol. 66, (1971), 622-626.

[292] Reiter, S. and Sherman, G., Discrete optimizing, <u>J. S.I.A.M.</u>, 13, (1965), 864-889.

[293] Rescigno, A. and Maccacaro, G. A., The information content of biological classifications, in C. Cherry (ed.), Information Theory, 4th London Symposium, Butterworths, London, (1961), 437-446.

[294] Rogers, D. J. and Tanimoto, T. T., A computer program for classifying plants, <u>Science</u>, Vol. 132, (Oct. 21, 1960), 1115-1118.

[295] Rohlf, F. J., Adaptive hierarchical clustering schemes, <u>Systematic Zoology</u>, Vol. 19, No. 1, (1970), 58-83.

[296] Rohlf, F. J. and Sokal, R. R., Coefficient of correlation and distance in numerical taxonomy, Kansas University Sci. Bull., 45, (1965), 3-27.

[297] Rose, M. J., Classification of a set of elements, Comput. J., Vol. 7, (1964), p. 208.

[298] Rosenblatt, M., Remarks on some nonparametric estimates of a density function, AMS 27, (1956), 832-837.

[299] Ross, G. J. S., Algorithms AS 13-15. Appl. Statist., 18, (1969), 103-110.

[300] Rota, G., The number of partions of a set, Amer. Math. Monthly, Vol. 71, No. 5, 498-504.

[301] Rubin, J., Optimal classification into groups: An approach for solving the taxonomy problem, J. Theoret. Biol., (1967), 15, 103-144.

[302] Ruspini, E. H., A new approach to clustering, Information and Control, Vol. 15, (1969), 22-32.

[303] Russell, P. F. and Rao, T. R., On habitat and association of species of anopheline larvae in South-eastern Madras, J. Malor Inst. India 3, (1940), 153-178.

[304] Sahler, W., A survey of distribution-free statistics based on distances between distribution functions, Metrika, Vol. 13, (1968), 149-169.

[305] Sammon, J. W., Interactive pattern analysis and classification, IEEE Trans. on Computers, Vol. C-19, No. 7, (July, 1970), 594-610.

[306] Sammon, J. W., Jr., On-Line Pattern Analysis and Recognition System (OLPARS), report number RADC-TR-68-263, Rome Air Development Center, Griffiss Air Force Base, New York, 72 pp., (1968).

[307] Samuel, E. and Bachi, R., Measures of Distances of Distribution Functions and Some Applications, Metron, Vol. 23, (Dec., 1964), 83-122.

[308] Sandon, F., The means of sections from a normal distribution, Brit. J. Statist. Psychol., Vol. XIV, Part II, (1961), 117-121.

[309] Sawrey, W. L., Keller, L. and Conger, J. J., An objective method of grouping profiles by distance functions and its relation to factor analysis, Educational and Psychological Measurement, Vol. 20, No. 4, (1960).

[310] Schnell, P., Eine methodl zur auffindung von gruppen, Biom. Z., Vol. 6, (1964), 47-48.

[311] Schweitzer, S. and Schweitzer, D. G., Comment on the Pearson
       r in random number and precise functional scale trans-
       formations, Amer. Sociol. Rev., Vol. 36, No. 3, (1971),
       517-518.

[312] Scott, A. J. and Symons, M. J., Clustering methods based
       on likelihood ratio criteria, Biometrics, Vol. 27,
       No. 2, (1971), 387-398.

[313] Scott, A. J. and Symons, M. J., On the Edwards and Cavalli-
       Sforza method of cluster analysis, Biometrics, Vol. 27,
       No. 1, (1971), 217-219.

[314] Sebestyen, George S., Automatic off-line multivariate data
       analysis, Proc. Fall Joint Comput. Conf., Spartan Books,
       (Nov., 1966), 685-694.

[315] Sebestyen, G. S., Pattern recognition by an adaptive process
       of sample set construciton, IRE Trans. on Info. Theory,
       Vol. IT-8, (Sept. 1962).

[316] Sebestyen, G. S., Recognition of membership in classes,
       IRE Trans. Info. Theor., (1961), IT-7(1), 48-50.

[317] Sebestyen, G. and Edie, J., An algorithm for nonparametric
       pattern recognition, IEEE Trans. Electronic Computers,
       Vol. EC-15, No. 6, (1966), 908-915.

[318] Sebestyen, G. S. and Edie, J., Pattern recognition research,
       Air Force Cambridge Res. Lab. Report 64-821, Bedford,
       Mass., (June 14, 1964).

[319] Shepherd, M. J. and Willmott, A. J., Cluster Analysis on
       the atlas computer, Comput. J., Vol. 11, (1968), 57-62.

[320] Shepard, R. N., The analysis of proximities: Multidimensional
       scaling with an unknown distance function, Psychometrika,
       27, (1962), 125-129, 219-246.

[321] Shepard, R. N. and Carroll, J. D., Parametric representation
       of nonlinear data structures, in P. R. Krishnaiah,
       Multivariate Analysis, Academic Press, New York, (1966).

[322] Sibson, R., A model for taxonomy II, Math. Biosci., 6, (1970),
       405-430.

[323] Sibson, R., Some observations of a paper by Lance and Williams,
       Comput. J., 14, (1971), 156-157.

[324] Silverman, J., A computer technique for clustering tasks,
       Technical Bull. STB 66-23, San Diego, Calif.: U.S. Naval
       Personnel Research Activity, (April, 1966).

[325] Singleton, R. C., Minimum squared-error clustering, Unpublished
       Internal Communication at Stanford Research Institute,
       Menlo Park, California, (1967).

[326] Smirnov, N. V., On the estimation of the discrepancy between empirical curves of distribution for two independent samples, Bull. Math. Univ. Moscow, Vol. 2, (1939), 3-14.

[327] Smith, J. W., The analysis of multiple signal data, IEEE Trans. on Information Theory, Vol. IT-10, No. 3, (July, 1964).

[328] Sneath, P. H. A., The application of computers to taxonomy, J. General Microbiology 17, (1957), 201-226.

[329] Sneath, P. H. A., A comparison of different clustering methods as applied to randomly spaced points, Classification Soc. Bull., 1, (1966), 2-18.

[330] Sneath, P. H. A., A method for curve seeking from scattered points, Comput. J., Vol. 8, (1966), p. 383.

[331] Sneath, P. H. A., Evaluation of clustering methods, in A. J. Cole (ed.), Numerical Taxonomy, Academic Press, London and New York, (1969), 257-267.

[332] Sonquist, J. A. and Morgan, J. N., The Detection of Interaction Effects, Survey Research Center, Institute for Social Research, University of Michigan, Ann Arbor, (1964).

[333] Sokal, R. R., Distances as a measure of taxonomic similarity, Systematic Zoology, 10, (1961), 70-79.

[334] Sokal, R. R. and Michener, C. D., A statistical method for evaluating systematic relationships, University of Kansas Sci. Bull., (March 20, 1958), 1409-1438.

[335] Sokal, R. R. and Rohlf, F. J., The comparison of dendrograms by objective methods, Taxon, Vol. 11, (1962), 33-40.

[336] Sokal, R. R. and Sneath, P. H. A., Principles of Numerical Taxonomy, San Francisco: W. H. Freeman and Company, (1963).

[337] Solomon, H., Numerical Taxonomy, Technical Report Number 167, Department of Statistics, Stanford University, 44 pp., (1970).

[338] Sorenson, T., A method of establishing groups of equal amplitude in plant sociology based on similarity of species content and its application to analyses of the vegetation on Danish commons, Biol. Skr. 5, (1968), 1-34.

[339] Spearman, C., Correlations of sums and differences, Brit. J. Psychol. 5, (1913), 417-426.

[340] Speilberg, K., On the fixed charge transporatation problem, Proc. 19th Natl. Conf., Association for Computing Machinery, 211 East 43rd Street, New York, (Aug., 1964).

[341] Spilker, J. J., Jr., Luby, D. D. and Lawhorn, R. D., Progress
       report--adaptive binary waveform detection, Tech. Report
       75, Communication Sciences Department, Philco Corp.,
       Palo Alto, Calif., (Dec., 1963).

[342] Srikantan, K. S., Canonical association between nominal
       measurements, J. Amer. Statist. Assoc., Vol. 65, No. 329,
       (1970), 284-292.

[343] Stanat, Donald F., Nonsupervised pattern recognition through
       the decomposition of probability functions, Tech.
       report, Sensory Intelligence Lab, Dept. of Psychology,
       University of Michigan, (April, 1966), 1-55.

[344] Stark, Lawrence, Okajima, Mitsuharu and Whipple, Gerald H.,
       Computer pattern recognition techniques: Electro-
       cardiographic diagnosis, Comm. ACM, 6, No. 10, (Oct.,
       1962), 527-532.

[345] Steinbuch, K. and Piske, U. A. W., Learning matrices and
       their applications, IEEE Trans. on Electronic Computers,
       Vol. EC-12, No. 6, (Dec., 1963).

[346] Stephenson, W., Correlating persons instead of tests,
       Character and Personality, Vol. 4, No. 1, (1935),
       17-24.

[347] Stephenson, W., The inverted factor technique, Brit. J.
       Psychol., Vol. 26, No. 4, (1938), 344-361.

[348] Stephenson, W., The Study of Behavior, University of Chicago
       Press, Chicago, Ill., (1953).

[349] Stewart, D. and Love, W., A general cononical correlation
       index, Psychol. Bull., Vol. 70, No. 3, (1968), 160-163.

[350] Stiles, H. E., The association factor in information re-
       trieval, Comm. ACM, Vol. 8, No. 1, (1961), 271-279.

[351] Stringer, P., Cluster analysis of nonverbal judgements of
       facial expressions, Brit. J. Math. Statist. Psychol.,
       (1967), 20(1), 71-79.

[352] Stuart, A., The correlation between variate values and ranks
       in samples from a continuous distribution, Brit. J.
       Statist. Psychol., Vol. VII, Part I, (1954), 37-44.

[353] Swain, P. H. and Fu, K. S., Nonparametric and linguistic
       approaches to pattern recognition, LARS Information
       Note 051970, Purdue University, Lafayette, Indiana,
       (June, 1970).

[354] Switzer, P., Statistical techniques in clustering and pattern
       recognition, Department of Statistics, Stanford Univ.,
       TR139.

[355] Tanimoto, T. T. and Loomis, R. G., A taxonomy program for the
       IBM 704, New York, International Business Machines
       Corporation (Data Systems Division, Mathematics and
       Applications Department), (1960), (M & A-6, the IBM
       Taxonomy Application.)

[356] Tatsuoka, M. M., The Relationship Between Canonical Cor-
       relation and Discriminant Analysis, and a Proposal for
       Utilizing Quantitative Data in Discriminant Analysis,
       Educational Research Corporation, Cambridge, Mass. (1955).

[357] Thomas, L. L., A cluster analysis of office operations,
       J. Appl. Psychol., (1952), 36, 238-242.

[358] Thompson, W. A., Jr., The problem of negative estimates of
       variance components, AMS, 33, (1962), 273-289.

[359] Thorndike, R. L., Who belongs in the family, Psychometrika,
       18 (1953), 267-276.

[360] Torgerson, W. S., Multidimensional scaling of similarity,
       Psychometrika, Vol. 30, No. 4, (1965), 379-393.

[361] Tryon, R. C., Cluster Analysis, Ann Arbor: Edwards Bros.,
       (1939).

[362] Tryon, R. C., Cluster analysis, Psychometrika, Vol. 22,
       No. 3, (Sept., 1957), 241-260.

[363] Tryon, R. C., Communality of a variable: Reformulation
       by cluster analysis, Psychometrika, 22, (1957), 241-260.

[364] Tryon, R. C., Comparative cluster analysis, Psychol. Bull.,
       36, (1939), 645-646.

[365] Tryon, R. C., Commulative communality cluster analysis,
       Educ. Psychol. Measmt., (1958), 18, 3-35.

[366] Tryon, R. C., Domain sampling formulation of cluster and
       factor analysis, Psychometrika, Vol. 24, No. 2, (June
       1959), 113-135.

[367] Tryon, R. C., General dimensions of individual differences:
       Cluster analysis versus multiple factor analysis, Educ.
       Psychol. Measmt., (1958), 18, 477-495.

[368] Tryon, R. C., Identification of social areas by cluster
       analysis, California: University of California Press,
       (1955).

[369] Tryon, R. C., Improved cluster-orthometrics, Psychol. Bull.,
       36, (1939), p. 529.

[370] Tryon, R. C. and Bailey, D. E., The BC TRY computer system
       of cluster and factor analysis, Multivar. Behav. Res.,
       (1966), 1, 95-111.

[371] Tryon, R. C. and Bailey, D. E., Cluster Analysis, McGraw-
       Hill Book Company, New York, (1970).

[372]   Tucker, Ledyard R., Cluster analysis and the search for
        structure underlying individual differences in psycholo-
        gical phenomena, Conference on Cluster Analysis of
        Multivariate Data, New Orleans, La., (Dec., 1966),
        10.01-10.17.

[373]   Turner, B. J., Cluster analysis of MSS remote sensor data,
        presented by Conference on Earth Resources Observation
        and Information Analysis Systems, Tullahoma, Tenn.,
        (Mar., 1972).

[374]   Turner, R. D., First-order experimental concept formation,
        Biological Prototypes and Synthetic Systems, E. E.
        Bernard and M. Kare, eds., Bionics Symposium 2, Ithaca,
        N. Y., Plenum Co., 1961.

[375]   Urbakh, V. Yu, A discriminate method of clustering, J.
        Multivariate Analysis, Vol. 2, (1972), 249-260.

[376]   Urbakh, V. Yu, On decomposition of statistical distributions
        deviating from normal into two normal distributions,
        Biofizika (Moskow), Vol. 6, (1961), 265-271.

[377]   Van Rijsbergen, C. J., A clustering algorithm, Comput. J.,
        13, (1970), 113-115, (algorithm 47).

[378]   Van Rijsbergen, C. J., A fast hierarchic clustering algorithm,
        Comput. J., 13, (1970), 324-326.

[379]   Vargo, L. G., Comment on "The Assignment of Numbers to Rank
        Order Categories", Amer. Sociol. Rev., Vol. 36, No. 3,
        (1971), 516-517.

[380]   Vinod, H. D., Integer programming and theory of grouping,
        JASA, (June, 1969), 506-519.

[381]   Von Mises, R., Wahrscheinlichkeitsrechnung, Leipzig-Wein,
        (1931).

[382]   Wacker, A. G. and Langrebe, D. A., Boundaries in MSS imaging
        by clustering, Proc. 9th IEEE Symposium on Adaptive
        Processes, (Dec., 1970).

[383]   Wacker, A. G. and Landgrebe, D. A., The minimum distance
        approach to classification, The Laboratory for Applica-
        tions of Remote Sensing Information Note 100771, Purdue
        University, Lafayette, Indiana, (Oct. 1971).

[384]   Wallace, C. S. and Boulton, B. M., An information measure
        for classification, Comput. J., Vol. 11, (1968), p. 185.

[385]   Watson, L., Williams, W. T. and Lance, G. N., Angiosperm
        Taxonomy: A comparative study of some novel numerical
        techniques, J. Linn. Soc., Vol. 59, (1966), p. 491.

[386]   Ward, J. H., Jr., Hierarchical grouping to maximize payoff,
         WADD-TN-61-29, Lackland AFB, Texas: Personnel Lab-
         oratory, Wright Air Development Division, (March, 1961).

[387]   Ward, J. H., Jr., Hierarchical grouping to optimize an
         objective function, J. Amer. Statist. Assoc., Vol. 58,
         No. 301, (1963), 236-244.

[388]   Ward, J. H., Jr., Hall, K. and Buchhorn, J., PERSUB Reference
         Manual, Report No. PRL-TR-67-3 (II), Personnel Research
         Laboratory, Lackland AFB, Texas, 60 pp., (1967).

[389]   Ward, J. H., Jr. and Hook, M. E., Applications of an hier-
         archical grouping procedure to a problem of grouping
         profiles, Educational and Psychological Measurement,
         Vol. 23, No. 1, (1963), 69-82.

[390]   Wherry, R. J., Jr. and Lane, N. E., The K-Coefficient, A
         Pearson-Type Substitute for the Contingency Coefficient,
         Report No. NSAM-929, Naval School of Aviation Medicine,
         Pensacola, Florida, 20 pp., (1965).

[391]   Wilks, S. S., Mathematical Statistics, John Wiley and Sons,
         Inc., New York, (1962).

[392]   Williams, E. J., Use of scores for the analysis of association
         in contingency tables, Biometrika, Vol. 39, Part 3/4,
         (1952), 274-289.

[393]   Williams, W. T., Dale, M. B. and Macnaughton-Smith, P.,
         An objective method of weighting in similarity analysis,
         Nature, Vol. 201, (1964), p. 426.

[394]   Williams, W. T. and Lambert, J. M., Multivariate methods in
         plant ecology I. Association analysis in plant com-
         munities, J. Ecology, Vol. 47, No. 1, (1959), 83-101.

[395]   Williams, W. T., Lambert, J. M. and Lance, G. N., Multi-
         variate Methods in Plant Ecology, J. Ecology, Vol. 54,
         p. 427.

[396]   Wishart, D., Mode analysis: A generalization of nearest
         neighbor which reduces chaining effects, in A. J. Cole
         (ed.), Numerical Taxonomy, Academic Press, New York,
         (1969), 282-319.

[397]   Wishart, D., An algorithm for hierarchical classifications,
         Biometrics, Vol. 22, No. 1, (1969), 165-170.

[398]   Wishart, D., FORTRAN II Programs for 8 Methods of Cluster
         Analysis (CLUSTAN I), Computer Contribution 38, State
         Geological Survey, The University of Kansas, Lawrence,
         (1969).

[399]   Wishart, D., A fortran II program for numerical classifica-
         tion, St. Andrew's University, Scotland, (1968).

137

[400] Wishart, D., Numerical classification method for deriving natural classes, Nature, London, (1969), 221, 97-98.

[401] Wolf, D. E., PROMENADE: Complete Listing of PROMENADE Programs, Appendix 9d to RADC-TR-68-572, Stanford Research Institute, Menlo Park, Calif., 465 pp., (1968).

[402] Wolfe, J. H., A computer program for the maximum likelihood analysis of types, Tech. Bulletin, 65-15, U. S. Naval Personnel Research Activity, San Diego, Calif., (May, 1965).

[403] Wolfe, John H., NORMIX-Computational methods for estimating the parameters of multivariate normal mixtures of distributions, Tech. Report, U.S. Naval Personnel Research Activity, San Diego, Calif., (Aug., 1967), 1-31.

[404] Wolfe, J. H., Pattern clustering by multivariate mixture analysis, Multivariate Behavioral Research, Vol. 5, No. 3, (1970), 329-350.

[405] Young, G., Factor analysis and the index of clustering, Psychometrika, Vol. 4, No. 3, (Sept., 1939).

[406] Yule, G. U., On measuring associations between attributes, J. Roy. Statist. Soc., Vol. 75, (1912), 579-642.

[407] Zadeh, L. A., Fuzzy sets, Information and Control, Vol. 8, (1965), 338-353.

[408] Zahn, C. T., Graph-theoretical methods for detecting and describing gestalt clusters, IEEE Trans. on Computers, Vol. C-20, No. 1, (1971), 68-86.

[409] Remote multispectral sensing in agriculture, Laboratory for Applications of Remote Sensing, Purdue University, Lafayette, Ind., Annual Report, Vol. 4, Research Bulletin 873, (Dec., 1970).

Vol. 59: J. A. Hanson, Growth in Open Economics. IV, 127 pages. 4°. 1971. DM 16,–

Vol. 60: H. Hauptmann, Schätz- und Kontrolltheorie in stetigen dynamischen Wirtschaftsmodellen. V, 104 Seiten. 4°. 1971. DM 16,–

Vol. 61: K. H. F. Meyer, Wartesysteme mit variabler Bearbeitungsrate. VII, 314 Seiten. 4°. 1971. DM 24,–

Vol. 62: W. Krelle u. G. Gabisch unter Mitarbeit von J. Burgermeister, Wachstumstheorie. VII, 223 Seiten. 4°. 1972. DM 20,–

Vol. 63: J. Kohlas, Monte Carlo Simulation im Operations Research. VI, 162 Seiten. 4°. 1972. DM 16,–

Vol. 64: P. Gessner u. K. Spremann, Optimierung in Funktionenräumen. IV, 120 Seiten. 4°. 1972. DM 16,–

Vol. 65: W. Everling, Exercises in Computer Systems Analysis. VIII, 184 pages. 4°. 1972. DM 18,–

Vol. 66: F. Bauer, P. Garabedian and D. Korn, Supercritical Wing Sections. V, 211 pages. 4°. 1972. DM 20,–

Vol. 67: I. V. Girsanov, Lectures on Mathematical Theory of Extremum Problems. V, 136 pages. 4°. 1972. DM 16,–

Vol. 68: J. Loeckx, Computability and Decidability. An Introduction for Students of Computer Science. VI, 76 pages. 4°. 1972. DM 16,–

Vol. 69: S. Ashour, Sequencing Theory. V, 133 pages. 4°. 1972. DM 16,–

Vol. 70: J. P. Brown, The Economic Effects of Floods. Investigations of a Stochastic Model of Rational Investment Behavior in the Face of Floods. V, 87 pages. 4°. 1972. DM 16,–

Vol. 71: R. Henn und O. Opitz, Konsum- und Produktionstheorie II. V, 134 Seiten. 4°. 1972. DM 16,–

Vol. 72: T. P. Bagchi and J. G. C. Templeton, Numerical Methods in Markov Chains and Bulk Queues. XI, 89 pages. 4°. 1972. DM 16,–

Vol. 73: H. Kiendl, Suboptimale Regler mit abschnittweise linearer Struktur. VI, 146 Seiten. 4°. 1972. DM 16,–

Vol. 74: F. Pokropp, Aggregation von Produktionsfunktionen. VI, 107 Seiten. 4°. 1972. DM 16,–

Vol. 75: GI-Gesellschaft für Informatik e. V. Bericht Nr. 3. 1. Fachtagung über Programmiersprachen · München, 9–11. März 1971. Herausgegeben im Auftag der Gesellschaft für Informatik von H. Langmaack und M. Paul. VII, 280 Seiten. 4°. 1972. DM 24,–

Vol. 76: G. Fandel, Optimale Entscheidung bei mehrfacher Zielsetzung. 121 Seiten. 4°. 1972. DM 16,–

Vol. 77: A. Auslender, Problemes de Minimax via l'Analyse Convexe et les Inégalités Variationnelles: Théorie et Algorithmes. VII, 132 pages. 4°. 1972. DM 16,–

Vol. 78: GI-Gesellschaft für Informatik e. V. 2. Jahrestagung, Karlsruhe, 2.–4. Oktober 1972. Herausgegeben im Auftrag der Gesellschaft für Informatik von P. Deussen. XI, 576 Seiten. 4°. 1973. DM 36,–

Vol. 79: A. Berman, Cones, Matrices and Mathematical Programming. V, 96 pages. 4°. 1973. DM 16,–

Vol. 80: International Seminar on Trends in Mathematical Modelling, Venice, 13–18 December 1971. Edited by N. Hawkes. VI, 288 pages. 4°. 1973. DM 24,–

Vol. 81: Advanced Course on Software Engineering. Edited by F. L. Bauer. XII, 545 pages. 4°. 1973. DM 32,–

Vol. 82: R. Saeks, Resolution Space, Operators and Systems. X, 267 pages. 4°. 1973. DM 22,–

Vol. 83: NTG/GI-Gesellschaft für Informatik, Nachrichtentechnische Gesellschaft. Fachtagung „Cognitive Verfahren und Systeme", Hamburg, 11.–13. April 1973. Herausgegeben im Auftrag der NTG/GI von Th. Einsele, W. Giloi und H.-H. Nagel. VIII, 373 Seiten. 4°. 1973. DM 28,–

Vol. 84: A. V. Balakrishnan, Stochastic Differential Systems I. Filtering and Control. A Function Space Approach. V, 252 pages. 4°. 1973. DM 22,–

Vol. 85: T. Page, Economics of Involuntary Transfers: A Unified Approach to Pollution and Congestion Externalities. XI, 159 pages. 4°. 1973. DM 18,–

Vol. 86: Symposium on the Theory of Scheduling and Its Applications. Edited by S. E. Elmaghraby. VIII, 437 pages. 4°. 1973. DM 32,–

Vol. 87: G. F. Newell, Approximate Stochastic Behavior of n-Server Service Systems with Large n. VIII, 118 pages. 4°. 1973. DM 16,–

Vol. 88: H. Steckhan, Güterströme in Netzen. VII, 134 Seiten. 4°. 1973. DM 16,–

Vol. 89: J. P. Wallace and A. Sherret, Estimation of Product. Attributes and Their Importances. V, 94 pages. 4°. 1973. DM 16,–

Vol. 90: J.-F. Richard, Posterior and Predictive Densities for Simultaneous Equation Models. VI, 226 pages. 4°. 1973. DM 20,–

Vol. 91: Th. Marschak and R. Selten, General Equilibrium with Price-Making Firms. XI, 246 pages. 4°. 1974. DM 22,–

Vol. 92: E. Dierker, Topological Methods in Walrasian Economics. IV, 130 pages. 4°. 1974. DM 16,–

Vol. 93: 4th IFAC/IFIP International Conference on Digital Computer Applications to Process Control, Zürich/Switzerland, March 19–22, 1974. Edited by M. Mansour and W. Schaufelberger. XVIII, 544 pages. 4°. 1974. DM 36,–

Vol. 94: 4th IFAC/IFIP International Conference on Digital Computer Applications to Process Control, Zürich/Switzerland, March 19–22, 1974. Edited by M. Mansour and W. Schaufelberger. XVIII, 546 pages. 4°. 1974. DM 36,–

Vol. 95: M. Zeleny, Linear Multiobjective Programming. XII, 220 pages. 4°. 1974. DM 20,–

Vol. 96: O. Moeschlin, Zur Theorie von Neumannscher Wachstumsmodelle. XI, 115 Seiten. 4°. 1974. DM 16,–

Vol. 97: G. Schmidt, Über die Stabilität des einfachen Bedienungskanals. VII, 147 Seiten. 4°. 1974. DM 16,–

Vol. 98: Mathematical Methods in Queueing Theory. Proceedings of a Conference at Western Michigan University, May 10–12, 1973. Edited by A. B. Clarke. VII, 374 pages. 4°. 1974. DM 28,–

Vol. 99: Production Theory. Edited by W. Eichhorn, R. Henn, O. Opitz, and R. W. Shephard. VIII, 386 pages. 4°. 1974. DM 32,–

Vol. 100: B. S. Duran and P. L. Odell, Cluster Analysis. A Survey. VI, 137. 4. 1974. DM 18,–